Preface

At first sight the transmission electron microscope may have seemed overwhelming. The prospect of learning to operate the instrument may also have been daunting. Similar feelings may now come to mind when considering maintaining the instrument. However, having mastered instrument operation, should not its maintenance be within your scope?

The basic techniques of electron microscope servicing are covered in this handbook. The procedures described will help you to analyse faults and in many cases solve the problem. I should add that, in the context of this handbook, the word 'problem' means 'anything from a slight loss in resolution, to a complete instrument breakdown'. An understanding of electron optics is assumed; refer to other handbooks or attend a transmission electron microscopy course if you feel weak in this direction.

It is impossible without years of training to become as proficient as a service engineer. Following the guide lines laid out in this book, be content to gradually develop your fault-finding technique as problems occur on your instrument. But start your maintenance career with the motto. . .

WHEN IN DOUBT DON'T

How should you approach fault finding? First follow Chapters 1 and 2 very carefully. In this way you will learn what to expect from the instrument when it is working correctly. Periodic reassessment will allow you to identify faults before they affect day to day microscopy.

When a problem does occur, working through Chapters 5 and 6 will help you isolate the area or system causing the trouble. Then you must ask yourself: 'Do I have the knowledge to go further, or should I call the engineer?' You have not failed if you call the engineer: you have failed if you pull the column apart and the fault is a dirty potentiometer!

What attitude should you take? A professional engineer would think 'as *man* built it *man* can fix it! In many cases simply taking a component apart, cleaning it, and reassembling will cure a fault. Ask the question: 'What was the last action prior to the fault occurring?' This may also throw light on the problem. Maintenance should damage neither the instrument nor the person carrying out the task. Double check every action before you carry it out. Look at the problem from every angle and do not rush. But when in doubt *don't*. Remember over-the-phone service is free, *ask*!

Finally it is easier to learn by watching someone else tackling a problem. Spend as much time as possible with your engineer and obtain a place on one of the electron microscope maintenance courses. Both will considerably advance your servicing capabilities and improve your confidence.

Acknowledgements

The author wishes to thank Professor T. Mulvey (Department of Mathematics and Physics, University of Aston in Birmingham) and Dr. N. G. Wrigley (National Institute for Medical Research, Mill Hill, London) for their contribution of Chapter 9 on the use of the optical diffractometer. Thanks are also due to Mr. M. J. Webb (National Vegetable Research Station, Wellesbourne) for Fig. 12, Mr. A. Nichells (Department of Metallurgy, University of Leeds) for Figs. 2 and 8 and Dr. D. J. Johnson (Department of Textile Industries, University of Leeds) for Fig. 5.

4.95

MICROSCOPY HANDBOOKS 08

Maintaining and monitoring the transmission electron microscope

S. K. Chapman
Protrain, Chinnor, Oxford

Oxford University Press · Royal Microscopical Society · 1986

Oxford University Press, Walton Street, Oxford EX2 6DP
Oxford New York Toronto
Delhi Bombay Calcutta Madras Karachi
Kuala Lumpur Singapore Hong Kong Tokyo
Nairobi Dar es Salaam Cape town
Melbourne Auckland
and associated companies in
Beirut Berlin Ibadan Nicosia

Oxford is a trade mark of Oxford University Press

Published in the United States
by Oxford University Press, New York

British Library Cataloguing in Publication Data
Chapman, S.K.
Maintaining and monitoring the transmission electron microscope. – (Microscopy handbooks; 08)
1. Electron microscope. Transmission
I. Title II. Series
502'.8'25 QH212.T7

ISBN 0-19-856407-4

Printed in Great Britain by
The Alden Press, Oxford

Contents

1 Standard conditions

The first maintenance task with any electron microscope is to monitor the characteristics of the instrument when it is performing well. Even if your microscope is serviced by the manufacturer it is better for you, and an aid for the servicing staff, if you have the figures relating to its basic performance.

1.1. The electron gun and high voltage system

These may be evaluated through the reading on the *emission current meter*. When the high voltage is switched on, the emission current meter will indicate the current passing between the high voltage system and earth. Even when the filament is not being heated the emission current meter will display this small **dark current**, a measure of leakage within the system.

Two types of beam current meter are in common use. One is usually calibrated from zero to 100 μA, the other has a higher current range.

With the first type of meter the current reading at each accelerating voltage is accumulated from zero. It is directly related to the emission current from the electron gun. The dark current at a specific accelerating voltage is the difference between the reading when the high voltage is switched off and when it is switched on with the filament current set at zero. If the high voltage system is in good condition, the dark current reading will vary from 0–5 μA, being highest at the maximum accelerating voltage.

The second type of meter does not record the true emission current for each accelerating voltage. At each voltage there is a fixed **standing current** which is a measurement of the current drawn by the high voltage system in generating the selected accelerating voltage. Included in this reading are the dark current and gun emission current. The manufacturer's instruction book will indicate the base current levels for the instrument. Switching on at each accelerating voltage, with the filament switched off, will indicate the standing current for that accelerating voltage. Any increase in this reading should be registered as the appropriate dark current. A fall in the level of standing current almost certainly indicates a drop in the high voltage level.

When first switching on the accelerating voltage there will nearly always be a surge in the emission current meter reading; this should settle within a few seconds. The higher the accelerating voltage the higher the surge and the longer the instability may last. If an instrument has not been used as its higher accelerating voltage levels for a considerable time, there may be an extensive settling period of up to an hour before the meter reading stabilizes.

High dark current and/or standing current readings indicate problems in the high voltage system: component breakdown within the high voltage tank, insulation breakdown in the high voltage cable or connections, a poor gun vacuum, or a dirty gun chamber and components. The problem area may be more accurately defined by following the procedures outlined in Section 6.5.

1.2. The emission system

This system may only be evaluated if the distance between the **filament** and the **grid cap (wehnelt cylinder)** is always set at a constant value. When fitting a new filament into the cathode assembly, the distance between the filament and the grid cap determines the maximum current that may be drawn from the high voltage circuit. If the **bias** or **emission** control is set at a constant value, short filament-to-cap distances increase the emission current compared with the manufacturer's recommended settings. Increase the filament-to-cap distance and lower current levels will be obtained with an increase in filament life. Less heat (a lower filament heating current) is required to reach saturation under these conditions.

At a specific accelerating voltage, with a specific filament-to-cap distance and at a specific emission current, the bias/emission control should always be in the same position if the system is trouble free. Under these same conditions, with a new filament the filament heating control should always be in the same position when the filament is *saturated.* As a filament ages and thins through oxidation and evaporation, the saturation point will be reached with a lower filament heating current. In order to determine these positions, set the zero point of the bias/emission and filament heating controls to an appropriate position, release the knob clamping screws, and adjust their alignment to the zero markings inscribed upon the desk.

1.3. The lens settings

The lens settings may be used to monitor the lenses themselves and the accelerating voltage level. A lens should always produce exactly the same action, i.e. have the same effect upon the electron beam, when its controls are set at the same positions. Any change in lens action at a specific accelerating voltage means a fault has developed within the associated electronics. If a similar fault is found with the other lenses either a common supply is at fault, or the high voltage level has changed. A rise in accelerating voltage would require a higher lens current to focus the beam, a drop in accelerating voltage would require a lower lens current to focus the beam.

If the instrument being evaluated has a **lens current meter** this should be used to monitor the maximum and minimum current values of each lens at each accelerating voltage. Other test conditions are required if the instrument does not have a lens current meter, or when assessing the accelerating voltage level: in each case control knob positions may be substituted for lens current values.

1.4. Condenser lens performance and high voltage level

Condenser lens performance and high voltage level may be related to the result of setting the *first condenser lens* (**spot size**) at a specific level and bringing the *second condenser lens* to **cross over**. Set condenser 1 at a specific number of steps from its minimum position. Increase the magnification to 10 000 × and bring condenser 2 to cross over – the point where the beam has reached a minimum diameter. Insert a small *condenser moveable aperture* to sharpen the spot and take a photograph. Measure the beam diameter on the developed negative and calculate the spot size in relation to condenser 1 and 2 lens currents or control positions:

$$\text{Spot size} = \frac{\text{Beam diameter (mm} \times 10^3)}{\text{Magnification}}\ \mu\text{m}. \qquad (1.1)$$

The factor 10^3 is simply the ratio between the measured units (mm) and the units in which the spot size is expressed (μm). If the spot size is to be expressed in nm the factor is 10^6.

1.5. Objective lens performance

Objective lens performance is critical when determining contrast level, resolution, magnification, and camera length. A periodic assessment of the objective lens is essential; an experienced operator will check the position of the focus controls after each specimen change. Figure 1 demonstrates the changes in performance corresponding to a change in **objective focal length**, stressing the need for constant focal conditions if performance is to remain constant.

1.6. Standardizing the objective focal length

This is essential during a specimen change and particularly prior to checking any part of the imaging lens system, the following procedure is used:

Ensure the specimen is flat, and placed into the specimen holder positioned so that the specimen on the grid sits facing away from the electron gun. Place the holder into the microscope and if the stage is **eucentric** adjust it to the **eucentric position**. Check to see that the specimen will focus at the control positions, or lens current value determined in Section 1.7.

1.7. Objective lens current

A routine check of objective lens current also constitutes a second check on the high voltage level. Insert a specimen and adjust the specimen as described in Section 1.6. Increase the magnification to 10 000 × and focus the specimen, using the **focus wobbler** if possible. A fixed magnification is important in order to standardize the field overlap that often occurs between the lens following the

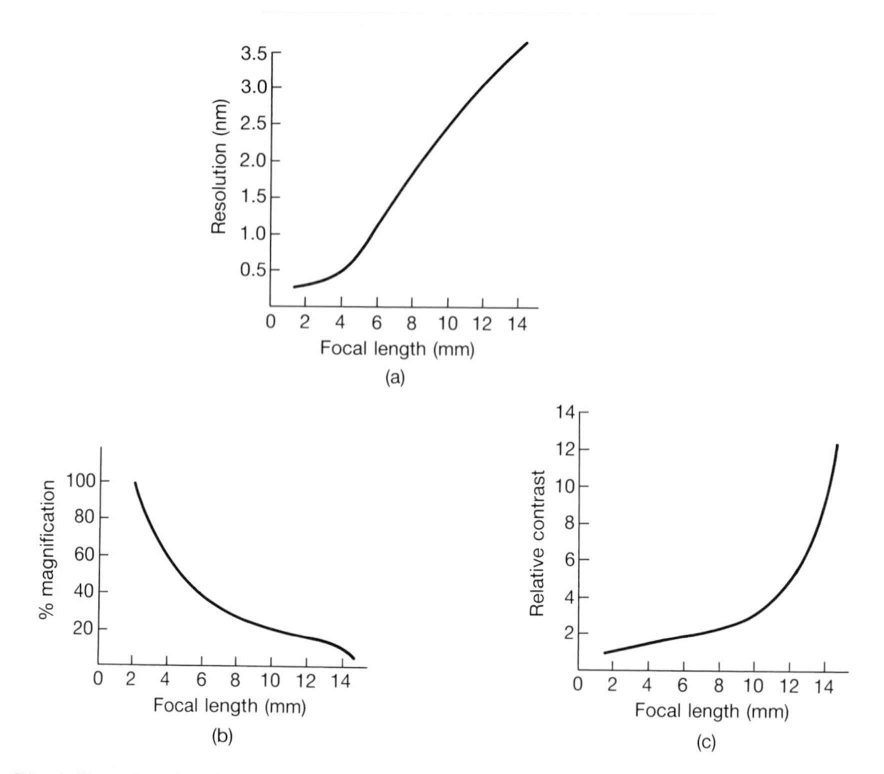

Fig. 1. Variations in objective lens performance related to focal length: (a) resolution, (b) magnification, (c) image contrast.

objective and the objective lens itself. At this point note the *objective lens control positions* and/or the *lens current*.

1.8. The objective lens stigmator

This should also be monitored. Most modern microscopes have lockable controls with numerical indications which simplify the recording of control strength. 'Old style' stigmators are best monitored by noting the positions of the controls. Remember that a stigmator is at zero when the X–Y controls are at their central position, two and a half turns on a five-turn potentiometer for example. Set the magnification to 50 000 to take a test reading. Do not use a ferromagnetic specimen as this itself will cause a field disturbance.

1.9. The intermediate or diffraction lens

This lens, which follows the objective, should be monitored as it will also act as a guide to the *high voltage level.* The introduction of a **diffraction** or **selected area**

Fig. 2. A caustic pattern.

aperture will restrict the field. In order to make the adjustment more precise, overfocus condenser 2. Record the lens current/knob position and the accelerating voltage used. This check finalizes the high voltage assessment, but it is also usable as an evaluation of **high voltage stability**. Set up the diffraction condition without a specimen, and a **caustic pattern** is produced (Fig. 2). Viewing the pattern through the instrument binoculars will display instability as a pulsing of the pattern.

1.10. Imaging lens combinations

Imaging lens combinations are complex in modern instruments, making a list of operating lenses and their lens currents at each magnification very useful (Fig. 3, Table 1). If a lens-current meter is not available the only alternative is to check the complete imaging system regularly using a magnification calibration standard. In some cases, if a graph of lens currents related to magnification is available, only a limited number of calibration points are needed to check the imaging system.

1.11. Magnification calibration

Magnification calibration is carried out either using a **line grating replica**, preferably a cross grating, or a **crystal lattice**, as follows.

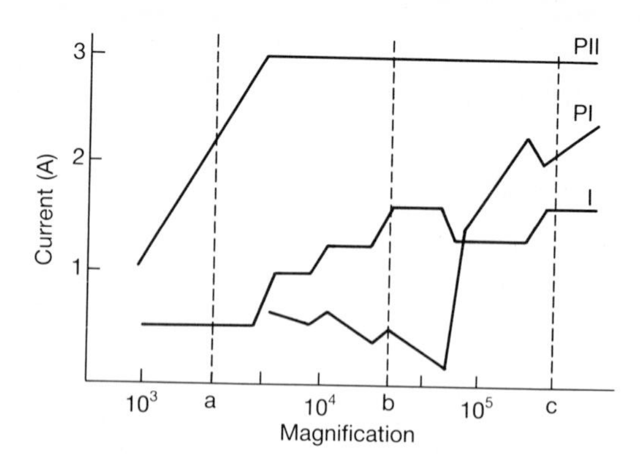

Fig. 3. Lens current relationships in a four lens imaging system (objective fixed). Points a, b and c, indicate test points which would together evaluate the magnification calibration. I stand for intermediate, PI for projector I, and PII for projector II.

Table 1. *Typical values of lens currents related to magnification*

		Lens currents (A)			
Control step	Indicated magnification	Objective	Intermediate (or diffraction)	Projector I (or intermediate)	Projector II (or projector)
1	1000	3.00	0.42	off	0.81
5	3000	2.98	0.40	off	1.48
10	8000	3.37	0.77	0.45	1.85
15	20000	3.47	0.87	0.35	2.39
20	45000	3.58	1.10	0.28	2.39
25	90000	3.53	0.95	1.00	2.39
30	200000	3.53	0.95	1.45	2.39
35	500000	3.61	1.13	1.75	2.39

(a) Place the specimen in the holder and adjust the instrument as described in Section 1.6.

(b) The focal settings must be normal, see Section 1.7, before proceeding.

(c) To reduce hysteresis effects in the lenses, approach each magnification from the same direction.

(d) Focus the image accurately and take photographs which incorporate the maximum number of calibration units, Fig. 4a.

(e) When only five grating units are included within the photographic area, look for and photograph an area containing a small defect, Fig. 4b.

(f) Calibration of the actual size of the defect will be made when the negatives are measured, the defect itself being used as the calibration unit for magnifications where less than five grating units have been measured, Fig. 4c.

(g) At magnifications in excess of 200 000 × the 0.344 nm (3.44 Å) *graphitized carbon lattice* is the best choice. This lattice should become visible on the screen at magnifications greater than 400 000 ×, Fig. 5. Over the calibration

Fig. 4. Micrographs of a cross grating of 0.4629 μm line spacing (2160 lines/mm), indicating use of a defect as a calibration unit at higher magnifications.

steps where the lattice is not visible, locate the lattice at a higher magnification, dropping to the desired level for focus adjustment and photography. Always approach the magnification to be calibrated from a lower magnification, paying particular attention when the area to be photographed has been found at a higher magnification. In other words, when reducing magnification, 'overshoot' and then move back up to the magnification to be calibrated.

(h) Measure the *negatives*, incorporating as many calibration units as possible. Always measure in two perpendicular directions as instruments may show a difference in magnification in one direction or another. Use a vernier scale for widespread structures and a calibrated hand lens for the finer structures.

The grating size is determined by dividing one by the number of squares/lines

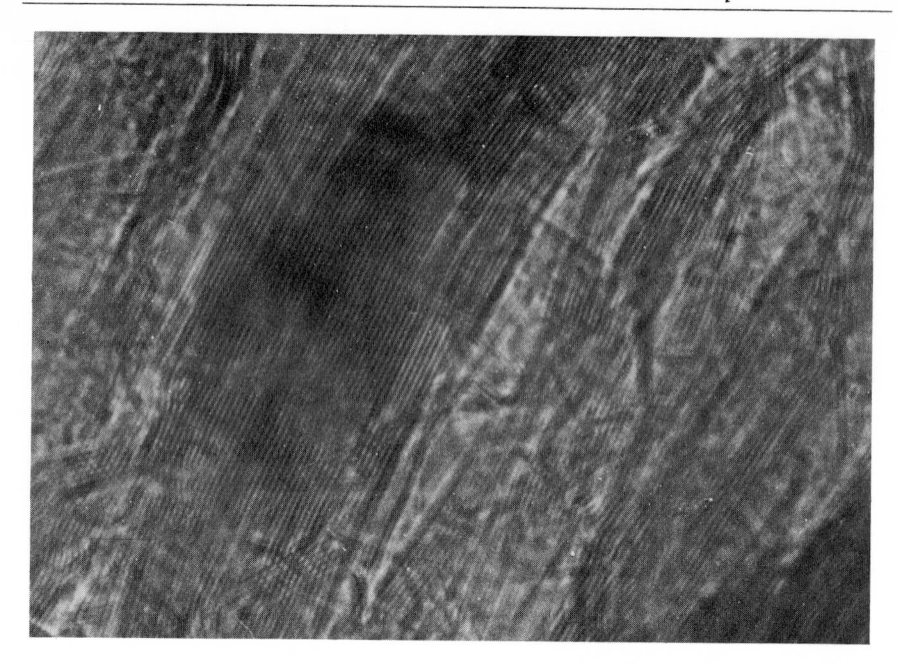

Fig. 5. Graphitized carbon lattice showing 0.344 nm spacing.

per millimetre, e.g. 2160 lines per millimetre grating has squares with $1/2160\,\text{mm} = 0.4629\,\mu\text{m}$ sides.

In a three lens imaging system plot intermediate lens current versus magnification, Fig. 6. In an imaging system using more than three lenses the magnification control is preprogrammed in order to simplify operation. In this case plot indicated magnification against the calibration magnification.

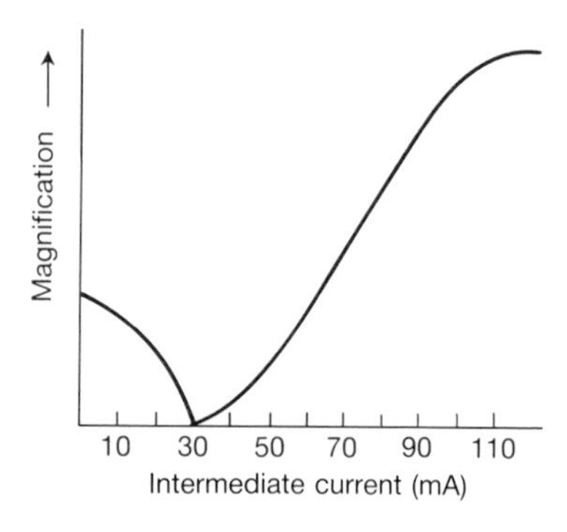

Fig. 6. Magnification vs. intermediate lens current for a three-lens imaging system at 100 kV (objective and projector lens currents fixed).

1.12 Magnification accuracy

Magnification accuracy in electron microscopy is beset with problems, as demonstrated in Table 2. The only accurate method for microstructural measurement is to include an internal standard with the specimen. Calibration of an instrument as described in Section 1.11 should only be used as a guide to magnification. Final figures should be rounded up or down but should not be used, for example, for accurate particle sizing (see also Section 9.2).

Table 2. *Sources of magnification errors*

Feature	Variation	Error
a. High voltage	± 1 kV change	1.5% error
b. Specimen position and objective lens strength	± 0.1 mm	3% error
c. Reproducibility of specimen position		2% error
d. Imaging lens strength change without recognition		> 10% error
e. Uni-directional calibration	distortion	2% error
f. Film (negative) stability	temperature/ humidity dependent	0.03% error
g. Paper (print) stability	Temperature/ humidity dependent	3% error
h. Resin coated paper (print) stability (e.g. Ilfospeed)		0.2% error

1.13. Electronics

In all instruments, valve or solid state, the electronics is similar in operation. A series of **standard supplies** are generated which are used to power the high voltage, lens, and deflection coil circuits. In valve circuits the standard supplies are very high, 100–700 volts, whilst solid state circuits use 5–72 volts. In earlier stabilizing circuits a standard **reference battery** or **reference supply** is used against which the circuit stabilities are matched. The reference portion of these circuits is very critical and is normally monitored through a check meter built into the instrument. All the standard values should be checked immediately general instability is detected; see Section 5.2.

Many modern instruments have comprehensive checking units which may be used to monitor critical areas of the electronics, e.g. standard supply voltages, lens circuit outputs, deflection coil circuit outputs, high voltage, and lens stabilities.

1.14. The vacuum system

You should also evaluate the vacuum system to obtain a set of **standard pumping times**. When the manufacturer fits **vacuum gauges**, monitoring performance is considerably simplified.

a. **Rotary pump performance** may be judged as the time taken for the pump to reach a particular vacuum level, 10^{-2} millibar for example. Let air into the pump prior to the test and repeat two or three times to obtain a standard time.

b. Time the *specimen exchange pump down cycle* to a specific vacuum level, 5×10^{-2} mbar for example.

c. Record the pump down time for an empty camera chamber.

d. Record the *gun airlock* pump down time.

e. Record the time taken for the *column* to pump down to a specific vacuum level, 10^{-5} mbar for example.

f. Another test may be possible on instruments without automated vacuum control; **leak back time**. This is the time taken for the column, with the camera chamber and pumping system isolated, to leak back to a specific level, 10^{-4}-mbar for example. Prior to this test the column must be pumped for at least four hours and liquid nitrogen traps attached to the column should be empty.

Most older microscopes have gauges calibrated in Torr. The SI unit of pressure is the Pascal (Pa) but most vacuum equipment manufacturers are still using the millibar (mbar) as an indication of vacuum pressure. One Pascal is equal to 100-mbar. To convert Torr to mbar, multiply by 1.33.

1.15 Vacuum level indicators

If an instrument is not fitted with vacuum gauges, standard conditions may still be monitored. In most instruments indicators are provided which light up when the area in question is evacuated to a level suitable for the diffusion pump to operate, and an additional light indicates when the vacuum level is suitable for the high voltage to be switched on. Checks 1.14b to 1.14e may be carried out using this information.

1.16. Standard conditions

Standard conditions should be recorded on a check list similar to that in Appendix 1, and used as a reference when problems occur. The details should be checked and corrected every six months.

2 Checking the performance of the microscope

There is a wide range of figures relating to an instrument's performance but in general a resolution test using a micrograph taken over a period of two to four seconds will suffice to evaluate resolution, high voltage and lens stabilities, astigmatism, drift, and contamination (see Chapter 9). It is not possible to take a 'good' resolution test photograph if any of these features are suspect, other than by reducing exposure time. Selective tests, evaluating only one of these features at a time, will aid the solution of problems in the future.

2.1. Resolution acceptance tests

Resolution acceptance tests performed by the manufacturer are a good guide to the overall performance of the instrument. The technique used is to demonstrate the resolution of a **crystal lattice** very near to the limit of the instrument's performance. Lattice resolution is a specialized technique which is not suitable for routine resolution evaluation; it only demonstrates that the instrument is capable or otherwise of reaching a particular resolution value. For day to day performance evaluation a **fresnel fringe** is a more suitable, and useful guide.

2.2. Fresnel fringe resolution

Fresnel fringe resolution tests are based upon the assumption that the smallest possible overfocus fringe width is very similar to the attainable point-to-point resolution. Fresnel fringes are observable at any point of large density change, with respect to electron scattering in the specimen, the edge of a hole being ideal.

In an underfocus condition the fringe is observed as a white band within the hole: the band reduces in width as focus is approached, the focus control being turned clockwise, Fig. 7a. At exact focus the fringe will be indiscernable. Turning the control to overfocus (continuing clockwise) the fringe is observed as a dark band which increases in width as the amount of overfocus increases, Fig. 7b. A photograph of the **overfocus fringe**, in a condition of overfocus which may not be visible when viewing with the binoculars, enables resolution levels down to 0.4 nm (4 Å), to be reached. The fringe test is valid up to several nanometres, the finest fringe attainable at any one time being the limit of the instrument's performance under the specific conditions being used.

Defects in the lens field due to **astigmatism** or **alignment errors** result in focus changes, which may be observed on photographs as a change in fringe thickness,

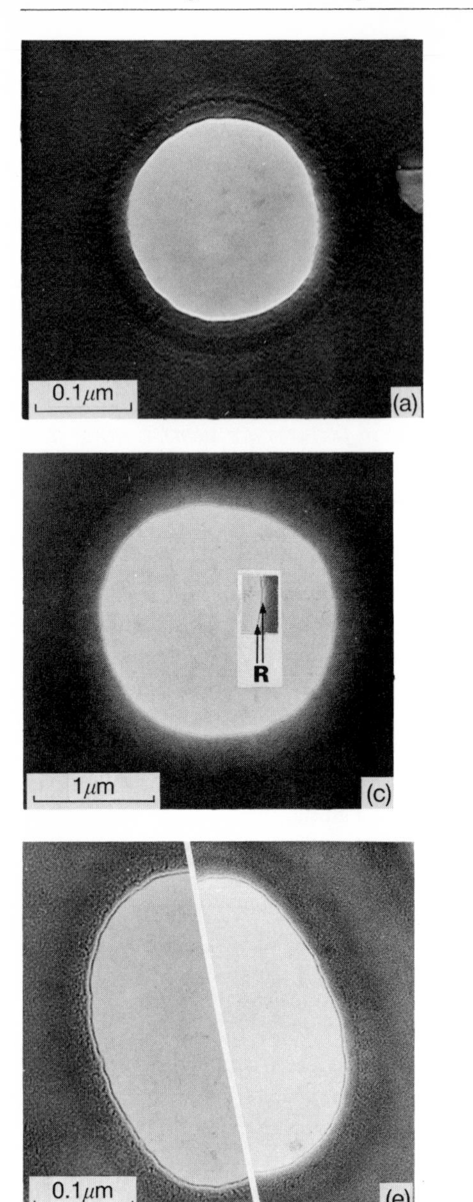

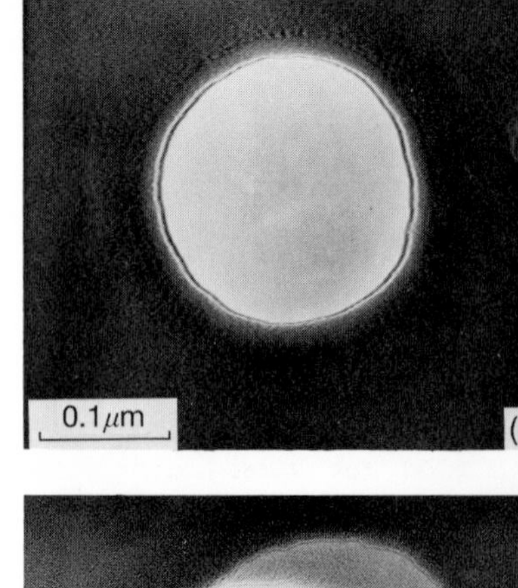

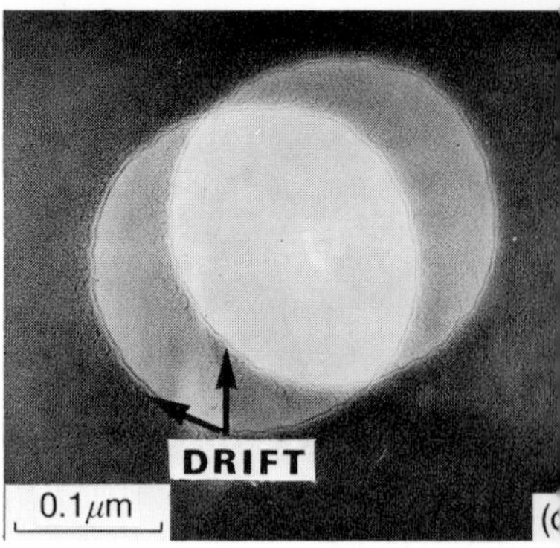

Fig. 7. Fresnel fringes in a holey carbon film.

(a) Astigmatic lens with strength greatest in the 'horizontal' direction. Underfocus fringes 'top' and 'bottom'.
(b) As (a), but with objective lens strength increased, showing overfocus fringes 'left' and right.
(c) Resolution test indicating (insert) the points of measurement.
(d) Drift rate test, 10 minute interval between micrographs.
(e) Contamination rate test, 10 minute interval between micrographs.

Figs. 7a and 7b. A full evaluation of Fresnel fringe photographs is made in Sections 5.11 to 5.13.

2.3. Performing a resolution test

(a) Insert a **holey carbon film** and adjust the microscope as described in Section 1.6.

(b) Set the magnification at 50 000 to 200 000 ×, whichever is the most appropriate for high resolution operation, having aligned the instrument (Chapter 4) to the best of your ability.

(c) Correct the astigmatism and study the overfocus fringe on part of the hole. The fringe width will almost certainly change as the high voltage or lens current drifts. Estimate the time involved for the focus to drift a number of fine focus steps. Your aim is to take a **through-focal series** of photographs, one underfocus, one in focus, and three just overfocus. This sequence increases the chance of obtaining an overfocus fringe suitable for measurement. Using this drift in focus, simply taking a series of pictures separated by a calculated time interval may produce the desired effect. If the drift in focus is small, check the number of focal steps between just underfocus and just overfocus. You will require to change focus four times during the test; divide the number of underfocus to overfocus steps by four to determine the number of steps between each photograph.

(d) Advance the film, check the focus, astigmatism, and image stability. You should only just be able to detect a very fine white fringe. Take the photograph slowly and steadily, without creating vibration.

(e) Advance to the next film, adjust the focal step (no need to view the image) wait a few seconds and take the photograph.

(f) Repeat 'e' until your series is complete.

(g) On viewing the image you should only just be able to see an overfocus fringe. If the fringe is very clear you have moved too far overfocus, the procedure should be repeated with smaller focal changes between the photographs.

Measure the negatives with a calibrated lens, from the centre of the black fringe to the centre of the white fringe as shown in Fig. 7c. Keep the negatives for reference, see Section 5.11.

$$\text{Resolution} = \frac{\text{Average fringe width (mm} \times 10^6)}{\text{Calibrated magnification}}\ \text{nm} \qquad (2.1)$$

Do not be surprised if you are unable to better 1 nm (10 Å) at your first attempt; practice is required.

2.4. Maximum performance

This will only be obtained if the instrument has stabilized and if special techniques are used. By taking more time over alignment and operation, an improved resolution limit should be achieved. The following modifications to your normal procedure will help in attaining maximum performance.

(a) Set the filament 0.2 mm closer to the grid cap, see Section 1.2, using a smaller aperture in the cap if this is available for your instrument. Align the instrument at the highest accelerating voltage, see Chapter 4.

(b) Use all of the anti-contamination devices during the whole of the stabilisation and test period.

(c) Insert a specimen and adjust the instrument as in Section 1.6, increase the magnification to your operating value, focus, and correct the astigmatism.

(d) Leave the instrument *on* under the above conditions for at least one hour.

(e) Select a condenser aperture half to two thirds the size of the aperture normally used and spot size of 1–3 μm.

(f) Overfocus the second condenser lens, turning the control (clockwise) away from cross over.

The steps taken in 'e' and 'f' will improve beam coherence. Overfocusing the second condenser moves the illumination crossover above the specimen. The cross-over point is the diameter of the selected spot size and it is this point which becomes a virtual source illuminating the specimen. Using smaller spot sizes or condenser aperatures and by moving even further overfocus with the second condenser continues to improve the beam coherence.

(g) Observe the image for at least 30 s to detect drift prior to taking the first photograph. Keep the room quiet during the work to prevent sonic vibration.

Most instruments use antivibration mounts to take account of floor vibration. The reduction of sonic vibration is very difficult to achieve, the act of limiting floor vibration may well make an instrument more sensitive to sonic frequencies.

2.5. Stage drift rate

Drift rate tests produce data which may be used to isolate this from similar problems and are carried out as follows.

(a) Use all anticontamination devices.

Liquid nitrogen traps on the diffusion pump, the main pumping manifold and in the form of a cold finger should always be kept filled during operation. If liquid nitrogen is not available a mixture of crushed solid carbon dioxide *and* acetone *will lower the temperature of these traps and offer some specimen protection.*

(b) Insert a holey carbon film and adjust the instrument as in Section 1.6. Set the stage as its central position.

(c) Set the magnification to 100 000 or 200 000 ×, whichever is preferable, focus, and correct the astigmatism on a hole about 10–20 mm across.

(d) Rock the stage drives very slightly without moving the specimen. This relaxes the drive system, taking the mechanical pressure off the components, and prevents the drive system from relaxing in its own time and spoiling the test.

This technique, which reduces stage drift, is always worth while with any form of high resolution microscopy.

(e) Set the focus to give a fine overfocus fringe and take your first picture using only two thirds of your normal exposure time. Retain the film for a double exposure. (On some instruments, e.g. JEOL 100C series, the film button should be depressed for the period between photographs to prevent the film advancing. An alternative would be to remove the camera drive fuse after the film is in position. This fuse is situated behind the door to the left of the camera, *but take care*!)

(f) Wait for exactly ten minutes with the instrument held in the photographic condition. Refocus the fringe to a similar width and take your second exposure, again at two thirds of your normal exposure time, Fig. 7d.

Measure the shift of one edge of the hole on the negative and calculate the drift rate from eqn. 2.2:

$$\text{Drift rate} = \frac{\text{Distance moved on the negative (mm} \times 10^6)}{\text{Magnification} \times \text{time (s)}} \text{ nm s}^{-1}. \qquad (2.2)$$

Typical drift rate figures should be between 0.2 and 0.01 nm s^{-1} depending upon the guaranteed performance of the instrument. The higher the guaranteed performance the lower will be the drift rate.

2.6. Contamination rate

Contamination, one of the most common causes of deterioration in image quality, is easily checked as follows:

(a) Insert a holey carbon film and adjust the instrument as in Section 1.6.

(b) Set the magnification to 100 000 or 200 000 ×, whichever is preferable, focus and correct the astigmatism on a hole about 20–30 mm across.

(c) Take your first photograph and start timing.

(d) Move the film and adjust the illumination to your normal viewing level.

(e) If you are using anti-contamination devices you will expect results in the region of 0.002 to 0.0005 nm s^{-1}; therefore a series of photographs should be taken, one every five minutes. If you are operating without anti-contamination devices a shorter time, perhaps two minutes, may be sufficient between exposures. Without anti-contamination devices the contamination rate will be up to 0.1 nm s^{-1}.

(f) Each picture is exposed with the normal low photographic intensity, the illumination being reset to observation levels between exposures.

(g) As the specimen and focus may drift they should be reset for each exposure.

Measure the diameter of the hole on each negative and calculate the contamination rate (nm s^{-1}) from the change in radius (mm) using eqn. 3. You may find that the contamination rate improves through your series as the specimen degasses.

2.7. Instrumental considerations

For an accurate contamination rate test a number of features should be considered. The test should not be made within four hours of switching the microscope on, or within a week of the column being disassembled. Both of these precautions allow time for the system to outgas and the vacuum to stabilize. The following instrument parameters affect contamination rate and should be logged with a test:

Accelerating voltage, emission current, spot size, condenser and objective apertures, number of film in the camera, and anti-contamination devices in use.

General operating procedures and cleanliness also influence the contamination rate, therefore:

Do not handle the area of the specimen holder that will be within the vacuum system during use.
Do not grease static 'O' rings.
Only very lightly grease moving seals.
Use anti-contamination devices at all times.
Pump the instrument 24 h a day if at all possible.
Outgass all photographic material prior to use in the microscope.

2.8. Camera length calibration

This may be used to accurately monitor high voltage fluctuations. Use a specimen made of an evaporated gold layer on a carbon or plastic film. The gold layer may be made by sputtering or by direct heating and a fairly heavy layer is required for 100 kV operation. Work in the following sequence:

(a) Insert the specimen and adjust the instrument as in Section 1.6, as a focal length error will also alter the camera length.

(b) Insert a selected area aperture which will reduce the number of crystals contributing to the pattern and sharpen the diffraction rings.

(c) Set the diffraction lens to form a diffraction pattern, and either adjust the **camera length** control or the **projector lenses** to an easily monitored position for future reference.

(d) Overfocus condenser 2 in order to improve the beam coherence and take a photograph with 20–40 s exposure.

Measure the ring spacing on the negative using a calibrated lens and calculate L from:

$$L = \frac{Dd}{2\lambda} \text{ mm} \tag{2.3}$$

Where L = camera length (mm), D = measured ring diameter (mm), d = lattice spacing (nm), and λ = wavelength of the electron beam (nm). The wave length of the electron beam, as related to the accelerating voltage, is as follows: 60 kV – 0.005 nm, 80 kV – 0.0042 nm, 100 kV – 0.0037 nm, 200 kV – 0.0025 nm, and 1000 kV – 0.00087 nm. Lattice spacings are indicated in Fig. 8. Since L is calculated in this test from an assumed value of λ, any variation in λ (ie, kV variation)

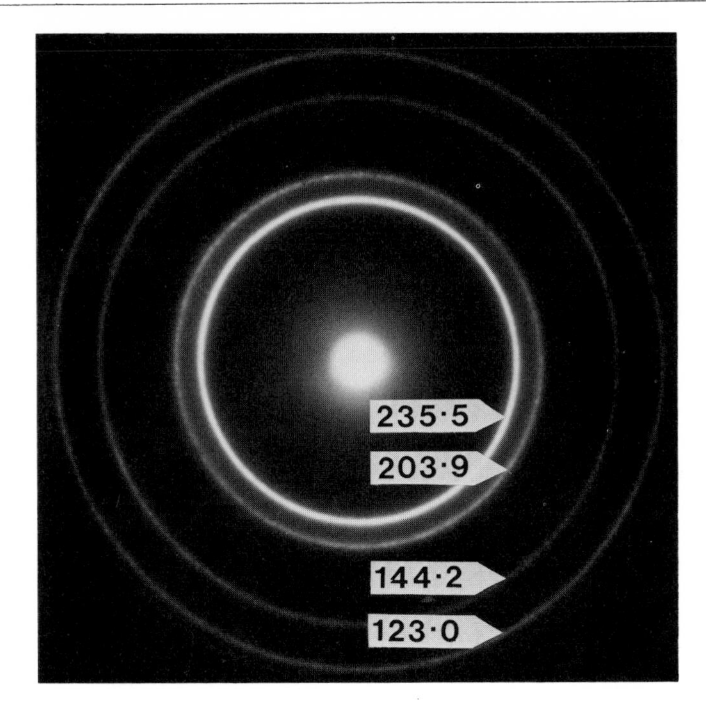

Fig. 8. Diffraction pattern of evaporated gold; *d*-spacings (pm) of the diffraction rings as indicated.

will give rise to a variation in ring diameter. An increase in ring diameter in comparative tests indicates a drop in accelerating voltage, a decrease in diameter indicates a rise in voltage. *Provided the operating parameters are kept constant this information is sufficient for a routine test.* Should the results of this test vary significantly consult the manufacturer for advice.

3 Cleaning techniques

There are two basic techniques for cleaning the column components of an electron microscope: by hand or with the aid of an ultrasonic cleaner.

3.1. Cleaning by hand

Cleaning the column elements by hand is by far the most fiddly method but it may often be the most rewarding. When cleaning small components by hand the technique must allow the cleaning medium to abrade every surface, nook, and cranny. Commercial sticks with cotton tips may be used for bulk cleaning. For small apertures and sharp corners 'home made' cotton sticks are superior; they also allow the use of cotton impregnated with metal polish. To produce a cotton tip a stick (orangewood or bamboo) is sharpened. A piece of cotton is teased out between the fingers until single strands are produced at one end. The fine strands are held against the tip and the stick spun between the fingers. This technique forms a cotton tip which is quite resilient and it enables tips to be formed of a size suitable for each cleaning task.

3.2. Cleaning media

These are available in several forms (brand names used)

(a) Liquid metal polish: Silvo, Brasso, Bluebell, etc.
(b) Wadding metal polish: Duraglit, Reiloreen, etc.
(c) Polishing paste: Wenol, Silver Foam (not long life), etc.
(d) Diamond compounds: (Particle size approximately 1 μm) Hyprez, etc.

All evidence of stains must be removed from both inside and outside the component. The pole pieces from any lens should be treated with great care. They should *not* be routinely cleaned and definitely not with any of the materials mentioned above other than Silver Foam. In extreme cases of contamination on a pole piece, the very light, water-soluble abrasive in Silver Foam is ideal.

3.3. Polish removal

Removing the polish after cleaning is as critical as the cleaning procedure. Originally engineers and microscopists used cotton sticks dipped in ether or acetone to wipe the metal polish clear, using a fresh face of the cotton stick for each wipe, and not dipping used tips in the fluid. This technique has fallen out of favour due to the vapour in the cleaning environment. If you are forced to use this technique, work in a well ventilated area.

An alternative technique is to wash the components clean using cotton swabs and hot soapy water. Baby bottle or test tube brushes are ideal for rigid screening tubes and for holes and threads in the grid cap. Count the components prior to and during the washing procedure in order to ensure that none are lost down drains or overflows. Dry each piece with an air or freon duster and wrap in Velin tissue prior to placing in a vacuum desiccator. The dried components may be removed from the desiccator, a camera chamber or a film desiccator, after about fifteen minutes pumping with a rotary pump.

3.4. Final checking

This should take place on a very clean bench covered in a sheet of clean white plastic. Use a chamois leather glove on your 'carrying' hand, whilst using an air or freon duster in your 'working' hand. Every piece should be checked very carefully with a 10 × hand lens just before placing it back into the microscope. If dirt is suspected, rub the area with Velin tissue to remove free debris, and blow air or freon through. Unclean components should be re-cleaned. If possible when replacing components in the column, always work from the top downwards to prevent debris falling on to cleaned components. Alternatively insert moveable apertures below the area on which you are working, as these will catch debris and are easily removed for cleaning. Always cover an open column with a clean blanking plate or Velin tissue.

3.5. Ultrasonic cleaning

Ultrasonic cleaning removes the personal touch from component cleaning but with very fine fragile pieces it is the only suitable method. Individual components are ultrasonically cleaned in a range of proprietary fluids: Quadralene, Inhibisol, or Arklone (all brand names). When dried, only inspection with a 10 × hand lens is required, as it is rare to find dirt remaining if the instructions on the use of the solutions are followed. A pre-clean by hand with metal polish may be required with severely contaminated cathode assemblies. If the external face of a component is critical, clean it alone; several components would rub together and may become marked. The cathode, anode, and specimen holders are particularly critical and should not be allowed to become scratched.

Cheaper alternatives may be preferred. Gun components that are made of stainless steel may be ultrasonically cleaned in a 5% ammonium hydroxide solution in water. This procedure will remove tungsten contamination, but more rigorous procedures are required for hydrocarbons. A low-cost cleaning procedure is listed below:

(a) Ultrasonically clean in 25% Silvo in water for 10 min or until the components are clean.

(b) Wash in running water until clear.

(c) Ultrasonically clean in 5% ammonium hydroxide in water for 1 min to remove the remains of the Silvo.

(d) Wash in running water until clear.

(e) Ultrasonically clean in two changes of alcohol, one minute each.

(f) Heat with a hot air blower or under a lamp.

(g) Check with a 10 × lens and replace.

3.6. The gun chamber

The gun chamber requires a totally different cleaning procedure from those mentioned earlier as it must be cleaned *in situ*. The gun area becomes dirty with the deposition of evaporated tungsten from the filament, and with a film of hydrocarbons from the vacuum system. The 'oily ozone' smell, which is characteristic of high voltage discharge, must be totally removed to ensure a stable system.

3.7. Dry cleaning

This may be used when the gun chamber is only lightly contaminated. If the chamber smells but is not visibly contaminated the smell may be removed by vigorously rubbing the chamber walls with a clean chamois leather. The insulator should be similarly treated if it has a glazed surface. A 10–20 min session should be sufficient to remove the smell and restore the system to maximum stability.

3.8. Wet cleaning

Severe chamber contamination will result in high voltage instability, the chamber will look dirty and smell strongly of oil and ozone. A microscope in this condition may only be cleaned with a set cleaning technique; to remove yellow or orange stains on the chamber walls, wipe in one sweep vertically down the side of the chamber using pieces of absorbent tissue or gauze damped with ether or acetone. Use each piece of tissue for a single wipe, take care to see that each wipe overlaps the previous wipe to prevent staining. Do not over-wet the tissue as runs will also cause stains. When the chamber has been wiped completely clean, follow the procedure on the insulator also *but only if it is glazed*. Check for and remove any missed stained areas. Dry clean the chamber as in Section 3.7, but in this case warming the chamber with a hot air blower will accelerate the process. Be careful not to heat the gun insulator beyond mildly warm or you will melt the vaseline filling which is used by some manufacturers. Cover the chamber when you are not working and insert the moveable condenser aperture to catch debris.

If the chamber has become stained dark orange, brown, or blue use metal polish. A wadding metal polish is best to remove heavy contamination, taking great care not to allow the liquid near to joints in the chamber construction. If the chamber consists of many units and therefore many joints, it will require removal from the microscope and total dismantling for major cleaning. After cleaning remove as much polish as possible with dry tissue or gauze, follow this with the wet and dry techniques described earlier in this chapter.

IF THE GUN AND, OR, CONDENSER SYSTEM HAVE BEEN COMPLETELY DIS-ASSEMBLED, ON REASSEMBLY THE MICROSCOPE SHOULD BE CHECKED FOR RADIATION LEAKS BY YOUR SAFETY OFFICER. THIS PROCEDURE APPLIES EVEN IF A PROFESSIONAL ENGINEER HAS PERFORMED THE SERVICE.

3.9. Platinum apertures

Platinum apertures may be cleaned in an alcohol or butane flame. Platinum tipped tweezers are used to hold the aperture about 3 mm above the blue part of the flame. Allow the aperture to reach orange-red heat for about 30 s; all of the surface will reach the same colour when it is completely clean. Repeat the process until the aperture is clean and then check it with an optical microscope. Check for cleanliness and aperture shape, rejecting mis-shaped apertures completely.

3.10. Molybdenum apertures

Molybdenum apertures must be cleaned in a **high vacuum** coating unit. Disc or thin strip apertures should be placed on a molybdenum boat, Fig. 9, bridging the gap between the two electrodes. Thick strip apertures may be clamped directly between the electrodes without using a supporting boat. At a pressure of lower than 10^{-4} mbar the apertures are heated to orange-red heat. This should be repeated until the aperture has no dark ring about the hole as it is heated. Overheating and multiple cleaning procedures tend to distort the aperture shape. Check the aperture with an optical microscope for cleanliness and aperture shape.

3.11. Thin film apertures

These fall into two categories being either made of gold and disposable, or of molybdenum and re-usable.

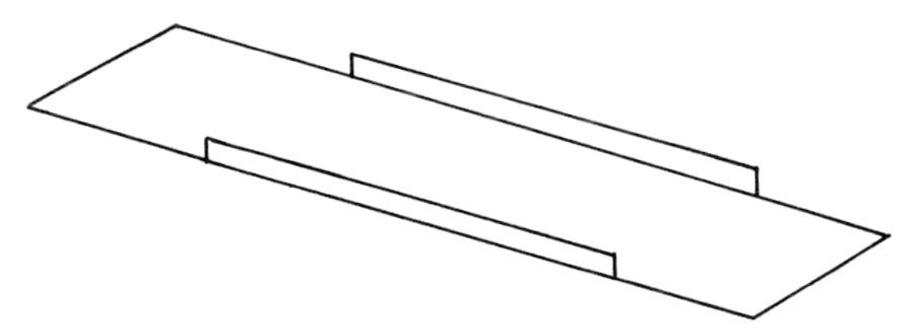

Fig. 9. Molybdenum foil boat for cleaning apertures.

If you do not use your microscope at 100 kV or above, the thin film apertures will not 'self clean' very efficiently. In this case leave the instrument running for a short period each week under conditions of high beam heating. Typical settings would be: 100 kV, emission 40 μA, spot size 15 μm, using each condenser and objective aperture for about five minutes. Concentrating the beam with condenser 2 and 'wiping' the edge of each objective aperture will accelerate the cleaning process. Molybdenum apertures may be cleaned as in Section 3.10, taking care not to overheat them.

4 Alignment

The TEM requires accurate alignment in order to simplify operation. When fault-finding, image defects may relate to alignment accuracy and therefore an understanding of the alignment is required.

A manufacturer will recommend particular alignment procedures for particular instruments. These procedures will be based upon general alignment principles, either from a condition when all the lenses are switched off, or final alignment procedures with all the lenses switched on. Set out below is a sequence of final alignment procedures that are suitable for attaining maximum performance and for fault diagnosis.

4.1. Electron gun alignment

This should be checked before inserting a specimen into the beam path. Image the source by bringing condenser 2 to cross over and magnify the image until it is about 50 mm in diameter. Lower the filament current, desaturating the filament, a spot and halo image will be formed. The halo should be symmetrical but lack of symmetry may be corrected with the **gun tilt** controls. If a correction is required retain the image on the screen centre with the **gun shift** controls (Fig. 10).

4.2. Saturation

Saturation is the condition where the electron gun is providing the maximum 'number of electrons' (beam current) for a given geometry. Visual indication of saturation is the convergence of the halo and the central spot when the source is imaged. The ideal situation is to heat the filament to the point where slight marks remain visible within the focused source. These will not be imaged during normal operation. You should remember that it is bad practice to operate with condenser 2 at cross over. This saturation technique is used by most TEM demonstrators, it improves filament life without loss of performance.

For optimum gun performance it is desirable to operate with the bias or emission control away from the fully clockwise position. Adjust the filament-to-grid-cap distance to obtain the desired emission current with the bias control as near to its mid point as possible. Variation in this setting results in changes in source size and emission efficiency.

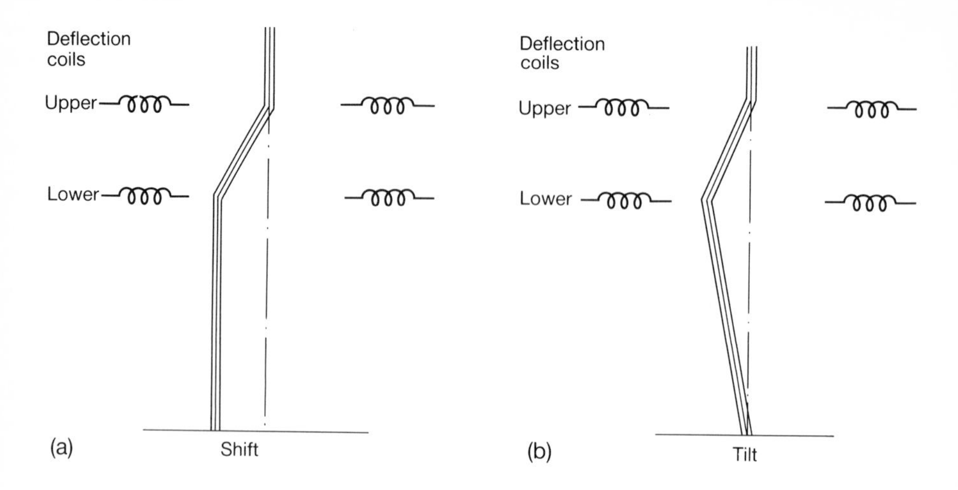

Fig. 10. Beam deflection systems: (a) shift or translate, (b) tilt.

4.3. Condenser apertures

These are supplied in a range of sizes. The largest aperture is intended for alignment and low magnification, the middle size aperture(s) for normal operation, and the smallest aperture for operating at high resolution.

To align an aperture, first check the illumination centre and correct any error with the **illumination shift**. Spread the illumination by overfocusing condenser 2 and recentre the beam using the **aperture alignment** controls. Bring condenser 2 back to cross over and recentre with the *illumination shift* controls. Repeat the procedure until a common centre is obtained. Do not underfocus condenser 2 or the alignment may require readjustment.

4.4. Balancing the beam deflection system

The beam should be kept near to the geometrical axis of the instrument. Unbalanced deflection, through too much emphasis being placed on the gun alignment coils, may result in the beam passing too near to the column wall. Any wall charge, created by the electrons striking contaminants on the wall, may distort the illumination.

The *gun* and *illumination deflection systems* are balanced as follows:

(a) Centre the illumination shift controls, for example a 5-turn potentiometer should be set to 2.5 turns.

(b) With a large spot size (10 μm) at about 1000× magnification, centre the illumination on the screen with the *gun* controls. Retain the correct spot and halo configuration.

(c) With a small spot size (2 μm), centre the illumination on the screen with the **illumination shift** controls.

(d) Repeat b and c until a constant centre is retained.

4.5. Condenser lens astigmatism

Condenser lens astigmatism is most easily corrected when imaging the source. Select the spot size which you intend to use for your investigation. Small spot sizes are desirable, for example use a spot size of about 2 μm provided there is sufficient illumination to work at 200 000 ×. Adjust condenser 2 to image the source at a magnification suitable to observe the striations within a slightly desaturated spot. Focus the striations (emission from extrusion markings on the filament) with condenser 2, and use the stigmator X–Y controls to further focus the image. Repeat the procedure until no further improvement is detected.

4.6. Objective lens alignment

The objective lens is aligned by setting its optical axis to coincide with those of the subsequent projector lenses, and to intersect the viewing screen at its centre point. In addition the illuminating beam should be centred along this axis.

Under these conditions, a specimen detail located on the optical axis will appear as an 'image point' at the centre of the screen and will not suffer lateral deviation as the focal length of the objective lens is altered slightly, for example by varying the lens current or the accelerating voltage. This image point is known as the 'current' and 'voltage' centre.

If a choice of *voltage* or *current* alignment is available maximum operating convenience is attained if the system is aligned on the *current centre* of the objective lens. After inserting a holey carbon film into the instrument, adjust its position as in Section 1.6. Focus the image at 10 000 × and overfocus condenser 2 in order to evenly illuminate the screen. Switch on the *objective lens current wobbler* or rock the focus control through the focal point. Using *illumination tilt* controls place the centre of the image movement at the screen centre. Repeat the procedure in steps of increasing magnitude to double your working magnification. Keep the illumination centred with the *illumination shift* controls.

An instrument with a *high voltage wobbler* enables you to align the objective lens to the *high voltage centre*, this is desirable when working with thick specimens or at high resolution. A similar technique to current alignment, it is the centre of the high voltage pulse that is placed on the screen centre.

The voltage and current centres are usually within 3 to 5 μm of each other but they have their own particular characteristics when fault finding.

If an instrument is accurately aligned for the objective current centre, high voltage instability will be exhibited as a directional blur. Depending upon its magnitude, objective lens instability may either be undetectable or the cause of general image softness.

If an instrument is accurately voltage aligned, depending upon the magnitude of the problem high voltage instability may either be undetectable or in the form of general softness. In this case objective lens instability would cause a directional blur. (See Chapter 5 for further details.)

4.7. Objective apertures

These are not always used when evaluating the instrument. A test is often simplified if the aperture is absent thereby excluding one complication, aperture cleanliness, from the investigation.

If an objective aperture is used it must be accurately aligned and this may only be achieved whilst observing the diffraction pattern, centring the aperture around the undiffracted centre spot with a specimen inserted.

4.8 Objective astigmatism

Objective astigmatism must be corrected at double the working magnification during test microscopy. All modern instruments use an X–Y stigmator system where the technique requires focusing the image and then using the X and Y adjustment like **fine focus** controls. The full procedure of focusing and then using the stigmators to improve focus is repeated until no further improvement in the image results. Do not look at the overall image, concentrate on one small area during the adjustment and do not select an area which is directional as this leads to over correction in one direction.

4.9. Imaging lenses

These must be aligned so that above 5000 × there is a constant image centre. Consult your manufacturer's manual to determine the procedures for your instrument. For example the large Hitachi (HU12A, H500), JEOL (100C series), and Zeiss (EM10) instruments mechanically align the final lens whilst ISI, Philips, the smaller Hitachi (HS9), and JEOL (100S series) instruments use electrical beam deflection. In the latter cases the controls are often placed behind panels or on the beam deflection circuit board.

4.10 Compensation devices

Compensation devices are built into most beam deflection systems. They may be mounted on a subpanel or on the alignment or stigmator circuit boards.

(a) Gun tilt correction is found on most ISI and JEOL instruments. These potentiometers allow compensation for tilt adjustment, retaining the illumination centre when the gun tilt controls are used. This action alters the balance of deflection between the two sets of coils. In balance the beam is retained in position whilst the tilt angle is adjusted. Compensation is carried out with X and Y controls being used to retain the beam on axis as the tilt controls are varied.

(b) Stigmator alignment prevents image movement during astigmatism correction. This adjustment is usually carried out using potentiometers mounted on the beam deflection or stigmator circuit boards. Each stigmator control (condenser and objective separately) is varied over its range with the illumination (condenser), or image (objective), held on axis through the X and Y compensation controls. The procedure is repeated for each stigmator X and Y control.

(c) Beam tilt compensation is provided to limit illumination shift whilst the illumination is being tilted. This feature is particularly important when carrying out beam tilt for dark field microscopy. Two sets of X and Y compensation controls are used to retain the beam centre over the tilt range.

(d) Illumination shift compensation is provided on some JEOL instruments. The field from the objective lens may overlap into the condenser lens system. This results in a 'total field' variation when the objective lens current is changed. This field change may cause illumination deflection which is compensated on JEOL instruments by an auxillary deflection coil mounted between the objective lens and condenser 2. Microscopes with mechanical alignment of the lens pole pieces compensate by optimizing the pole piece positions.

(e) High voltage compensation is provided on small JEOL instruments. These controls are mounted on the high voltage circuit board and are used to centre the image at each kV.

(f) Zoom compensation is amounted within the objective lens circuit of Hitachi and JEOL instruments. This allows the focus change at each magnification to be minimized by varying the focal length through the use of jump leads or potentiometers.

(g) Wobbler compensation is usually mounted on a sub-panel or on the beam deflection circuit board. The system allows adjustment of the beam to form a single spot at the maximum wobbler strength. Adjustment of the wobbler may affect the illumination tilt compensation as the same coils are used in both cases.

(h) Scan centre (low magnification centre) is often mounted on the beam deflection circuit board. These controls centre the 'normal' field of view in the scan mode. This correction is required as the objective lens alignment alters at the very low current range used.

5 Fault finding

Many of the problems that arise during the routine operation of a transmission electron microscope may be solved very simply. The only stipulation is to follow a very strict procedure, assessing the situation one step at a time.

5.1. First procedures

Faults seldom 'suddenly occur', most are operator induced and checking the previous procedure often leads to an immediate solution: e.g.

Instability : Specimen holder not correctly seated.

No beam : Filament broken, objective aperture out of alignment, or the camera shutter has not reset.

Secondly examine the controls: are they in their recognized *standard positions*; see Appendix 1. Do not touch anything – just look! If the problem is associated with the image, study the specimen detail through the binoculars.

5.2. Specimen observation

Is the specimen itself causing the problem? If you observe a steady drift in one direction this is either specimen or stage movement. Traverse the specimen to a position where both the specimen and a grid bar are in view. Watch the image and vary condenser 2. If the specimen is moving on the grid, it will move while the grid bar stays still. If it is the grid that moves, the specimen clamp may not be sitting correctly; the specimen holder clamping surfaces may be dirty; or the seat within the stage, which holds the specimen holder, may be dirty or damaged.

To finally isolate the fault to the specimen, the holder, or the stage, insert a 'thermally stable' holey carbon film and check its stability. If the image remains unstable use a different holder; if the instability remains, the fault must be in the stage itself. Check how the drift compares with your standard drift rate by taking a test picture, see Section 2.5.

If after observing the image, the magnification or illumination does not seem to be normal, it could be that a lens is no longer operating. Check your lens current readings and other test values. At this stage check to see if the problem occurs at another kV. If the fault is restricted to one or two kV positions it could be due to a faulty relay or switching problem. If the magnification is suspect check it against your own calibration figures, see Section 1.11. More complex problems such as difficulty in focusing or in correcting astigmatism may only be resolved by taking test pictures.

5.3. Investigative microscopy

Investigative microscopy involves the microscope itself becoming your test instrument through a series of photographs. These test photographs are taken in the same way as those for a resolution test. In this case the five photographs should be taken over a range of focus settings in order to produce two underfocus pictures, one in focus, and two overfocus. Use the photographic procedure outlined in Section 2.3, under the same alignment conditions as when the fault was first noticed and at the same kV. Repeat the procedure at a different kV, as problems which are removed by a kV change may be related to charge–discharge (see Section 5.8), or a component (relay or switch) only used at one accelerating voltage. Aligning an instrument, focusing, and correcting astigmatism may be very difficult with certain defects, such as high voltage or lens instabilities.

5.4. Total instrument failure

If the instrument will not operate even more basic checks must be made.

(a) Are all power switches on; is the main fuse in the power line broken?

(b) Is the main fuse on the instrument broken?

(c) If compressed air is used to operate the vacuum valves, is the pressure correct?

(d) Are the correct water flow and temperature being maintained?

(e) Has the rotary pump fuse or drive belt broken?

(f) Have all the connections between the instrument and the power unit or the power supply been checked? During routine laboratory cleaning these connections are often disturbed.

5.5. No high vacuum

If the instrument will not switch to high vacuum but has moved into a low vacuum cycle check:

(a) The water flow and temperature are correct.

(b) The diffusion pump is warming up: listen for the tinkling sound made by the fluid in the hot pump, or remove a suitable panel, probably directly behind the column, and *with great care* hold your hand *near* to the base of the pump. The pump should feel very hot within 20 to 30 minutes of switching on. If the pump is cold, check the diffusion pump fuse; if this is sound you can assume that the diffusion pump heater is broken. Confirm this through electrical investigation using the guide in Section 7.9.

(c) The vacuum level may not be sufficient for the hot diffusion pump to be opened to the column. A vacuum leak may be the cause, or the vacuum control circuit, or a vacuum gauge may have a fault.

5.6. Repeated low–high vacuum cycling

This may be caused by one of the following faults:

(a) A vacuum leak too great for the rotary pump backing the diffusion pump to cope with, see Section 7.5.

(b) Diffusion pump not working, see Section 5.5b.

(c) Compressed air-operated systems: when the reservoir pressure is close to its minimum, it drops below the minimum level when a number of valves operate simultaneously. The drop in pressure closes all of the valves until the correct pressure is regained, at which point the vacuum system will re-cycle.

5.7. Lenses keep switching off

If the lenses keep switching off but the rest of the microscope seems to be functioning correctly, check the water flow used to stabilize the temperature of the lens power supply units; when they overheat they switch off!

5.8. Condenser lens system

Being nearer to the electron gun than any other lenses in the instrument, the condenser lens system is heated by the passage of the electron beam to a much higher level than the imaging system, resulting in contamination of the pole pieces and/or their screening tubes. Problems in the condenser lens system will either be electrical, or due to the build up of contamination, and may be observed by imaging the illuminating spot, i.e. focusing condenser 2. A dirt particle will be non-conducting and therefore charge when the electron beam falls upon it. The charge will build up to a point where the potential is sufficient to bridge the charge to ground. Whilst the beam continues to fall upon the particle this process of charge and discharge is repeated. During the charging process the beam is deflected and on discharging rapidly moves back to its original position; this is demonstrated in Fig. 11. Decreasing the kV or increasing the emission current will increase the charging rate. If the dirt is below the aperture position, fitting a smaller condenser movable aperture reduces the rates of charge; if no change in the rate of charge is observed the dirt must be above the aperture.

5.9. Specimen area

Dirt in the specimen are will cause problems in one of two ways:

(a) If the dirt is above the specimen, a charge–discharge problem will occur as in the condenser system.

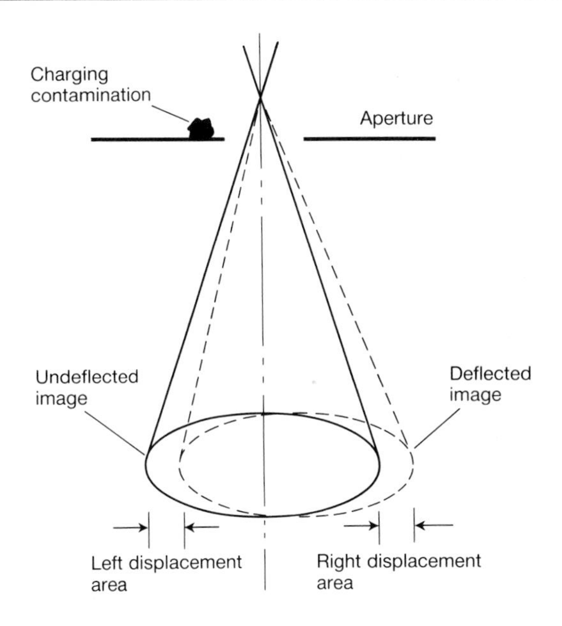

Fig. 11. Beam shift associated with charge/discharge at a dirt particle.

(b) If the dirt is below the specimen, the image astigmatism will be increased.

Reducing the kV, increasing the emission current, using a larger condenser aperture, or using a larger spot size will all increase the problems caused by dirt in this area. If the dirt is situated just below the objective aperture its effect may be decreased if the smaller objective aperture is used.

In an emergency, but only as a temporary measure, using a higher kV may reduce the effect of dirt in the specimen area sufficiently to obtain acceptable results.

5.10. Day to day micrographs

'Day to day' micrographs may often be used to diagnose instrument problems. Some faults that may be observed are broken down into the following categories (see also Chapter 9) and related to the examples in Fig. 12.

(a) **Drift**, Fig. 12a: a general streaking of the image in one direction.

(b) **Astigmatism**, Fig. 12b: a minor streaking of the image in one direction, often only visible directly in micrographs above 50 000 ×.

(c) **Contamination**, Fig. 12c: darkening and softening of the image which relates to the time the beam has been on the field of view prior to photography; the longer the time the softer the image. 'Softness' is very difficult to describe, but the image is not as contrasty or sharp as one would expect, appearing as if viewed through a hazy filter.

(d) **Vibration**, Fig. 12d: the image may vary from being blurred to multiple

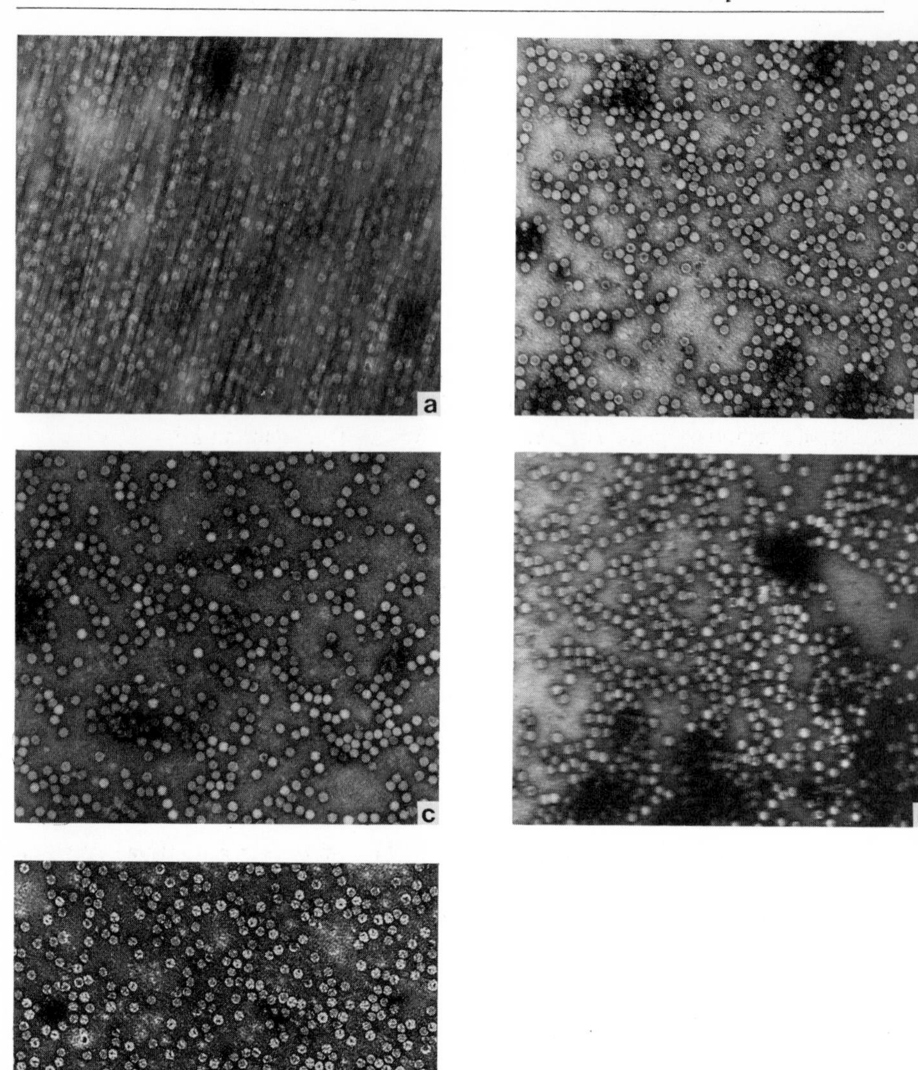

Fig. 12. Micrographs of Tomato Bushey Stunt virus (particles 28–30 nm diameter)
(a) Specimen drift.
(b) Astigmatism.
(c) Contamination.
(d) Vibration.
(e) Out of focus.

images as the magnitude of the vibration increases. High frequency vibration in one direction may be confused with astigmatism or drift.

(e) **Out of focus**, Fig. 12e: a general softness of the image; high contrast areas are retained but the image is not sharp.

Whilst these evaluations of normal micrographs may help with the diagnosis of random errors in operation, a true test micrograph (section 5.3) is invaluable when problems persist.

5.11. Test micrograph evaluation

The results of Section 5.3 rely upon an evaluation of the Fresnel fringe on a negative using a 10× hand lens. An overview of all five negatives from one series may allow a simple diagnosis, when compared with the results from Section 2.3.

(a) A problem above the specimen usually results in an enlarged source; the loss of coherence produces fringes of low contrast in an 'out of focus' condition, while the 'in focus' image will look quite sharp.

Source – circuit instability in the high voltage, condenser lenses, or beam deflection coils; alternatively charge–discharge due to dirt in these areas.

(b) A problem below the specimen results in a general degradation, both 'in focus' and 'out of focus' fringes show poor contrast.

Source – circuit instability in the imaging lenses, AC or switching DC fields from external sources or vibration.

5.12 Overfocus fringe defects

Overfocus fringe defects may be related to the image forms displayed diagrammatically in Fig. 13.

(a) **Double image** (Fig. 13a): this may be caused by a very old filament which has become severely rounded producing multiple emission when slightly undersaturated. If the instrument is affected by a large AC field, or oscillating beam deflection system, a double image may be formed. Always re-check multiple image problems as they may be confused with AC field defects, i.e. check the filament saturation.

(b) **General softness** (Fig. 13b): this is most commonly caused by contamination. If the instrument is very accurately aligned a fault in the high voltage or lens

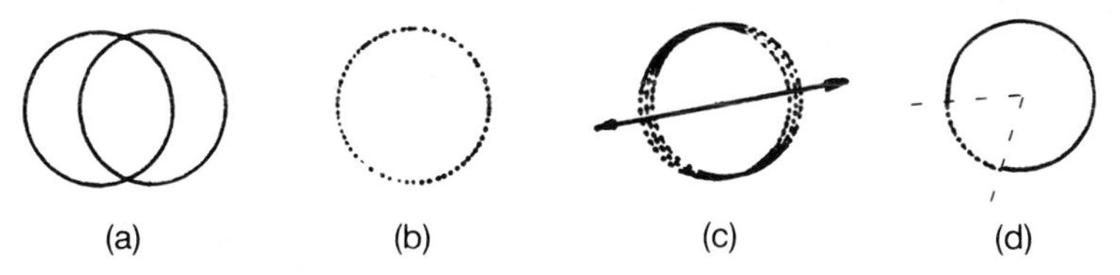

Fig. 13. Overfocus fringe defects in a holey carbon film:
(a) double image, (b) general softness, (c) directional softness, (d) vectorial softness.

circuits produces a unidirectional pulse which results in general softness. To isolate a problem of this kind, misalign the column step by step. Take test photographs as in Section 5.3 at each stage. Misaligning a lens exaggerates a fault, simplifying isolation of the lens which is causing the problem. Misalignment of illumination tilt will exaggerate high voltage instability (see Section 4.6). Lens problems may also be isolated by switching off each lens in turn. Take test photographs at each stage; these should be enlarged to give a constant magnification in order to check if the fault has been removed. General softness may be caused by too short an exposure during test microscopy, i.e. photographing an AC or DC problem and recording only part of a cycle. Figure 14 demonstrates how a four second exposure during test microscopy optimizes the information recorded.

(c) **Directional softness** (Fig. 13c): in its most common form this is due to astigmatism; however, vibration, magnetic fields, high voltage, and lens instabilities may all be manifested in this way. If the problem is due to a magnetic field it will decrease at a higher kV and increase at a lower kV. Field or deflection problems

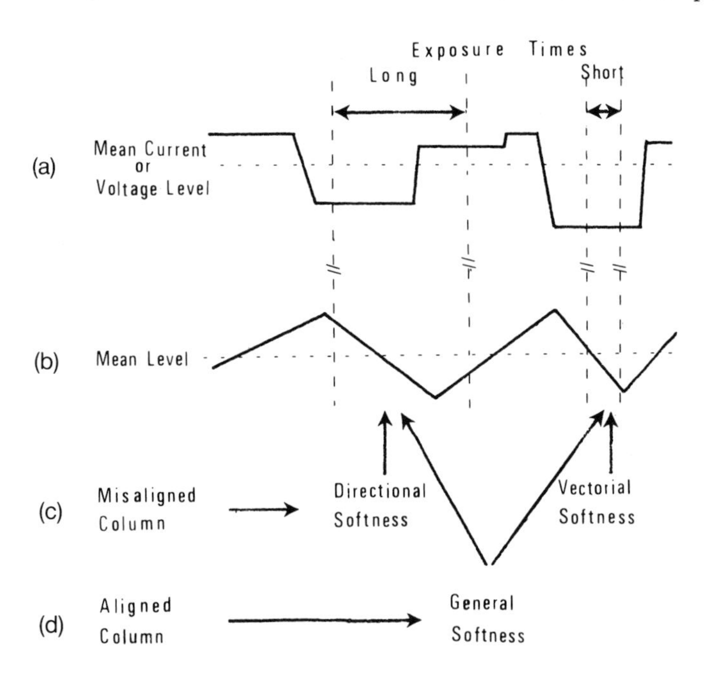

Fig. 14. The effect of exposure time on image quality.
(a) Demonstrates a pulsing instability or field effect, a condition where the deviation from a mean is a series of steps, the levels of which remain constant for a variable time span.
(b) Demonstrates a constantly changing instability or field effect, often characterized by a 'sawtooth' wave form.
(c) When the microscope colum is misaligned the defects shown in (a) and (b) will produce an image with 'directional softness' when a long (~ 4 s) exposure time is used and an image with 'vectorial softness' when a short (~ 1 s) exposure time is used.
(d) When the microscope column is accurately aligned with defects shown in (a) and (b) will produce an image with 'general softness' for both short and long exposure times.

within the microscope itself may be isolated by switching off deflection circuits one by one using mechanical alignment where possible. The direction of residual vibration may be deduced by striking the column whilst viewing the image or taking a photograph. This result may be compared with the image defect; mechanical vibration will always have a specific direction at the same magnification. Pulsing DC field or charging (Section 5.8) also produce a similar image defect. In both cases fault analysis through an increase in kV will show a reduction in the magnitude of the problem. However, charging is the only one showing an increase in the problem with an increase in beam current, i.e. a higher emission current or a larger condenser aperture size or spot size. Charging is removed by cleaning the instrument. An external magnetic field may be more difficult to isolate and must be investigated by trial and error unless a field coil and meter are available. As a last resort, physically turn the microscope through a 45–90°. If the direction of the defect remains the same in relation to the room, there *is* an external field.

(d) **Vectorial softness** (Fig. 13d): this is an image defect that affects only one quadrant of the test photograph and is often known as **quad**, it may be produced through a slight misalignment during the manufacture of a lens pole piece for the imaging system or by photographing a defect using too short an exposure time, Fig. 14.; check the objective lens alignment, see Section 4.6.

5.13. Fresnel fringe shape

The fresnel fringe shape may be used to analyse a defect which has been photographed during test microscopy.

(a) **Gradual taper** between two areas of fringe within a test photograph, Fig. 15a, suggests that the change in fringe width is due to astigmatism.

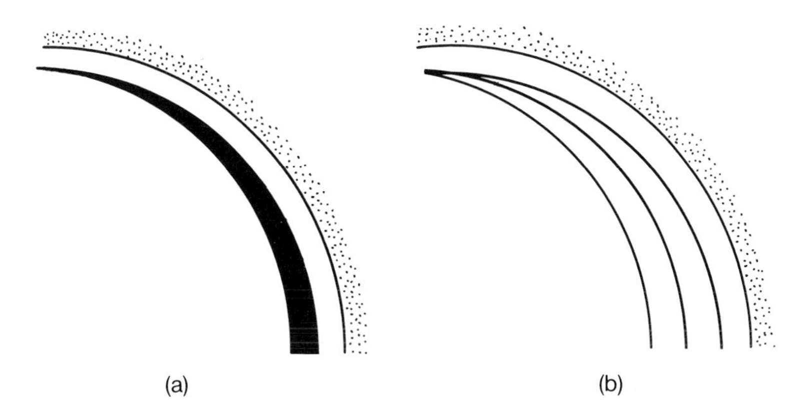

Fig. 15. Fresnel fringe shapes.
(a) Gradual taper, or variation in fresnel fringe width arising as a result of astigmatism.
(b) Sharp taper, or abrupt variation in fresnel fringe width arising as a result of an external AC field, lens or deflection coil current oscillation.

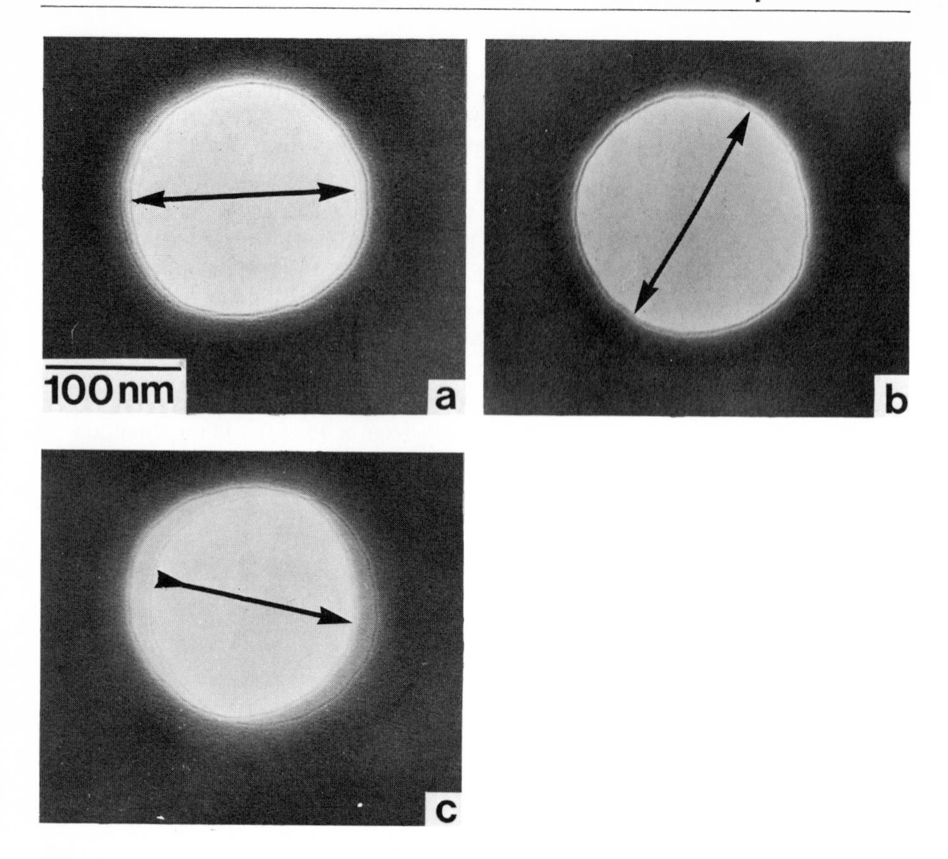

Fig. 16. Micrographs illustrating field effects or vibration (arrows indicate direction of field).
(a) A large AC field producing a triple fringe.
(b) A small AC field producing fringe 'softening'.
(c) A pulsing DC field giving a 'normal' and displaced image.

(b) **Abrupt changes** in the width of a frainge within a test photograph, Fig. 15b, indicates that the change in fringe width is due to a field problem.

(c) **AC fields or vibration** produce test photographs that are made up of two or three superimposed images. If the field is present throughout the exposure two images are formed, one at each end of the deflection. If the field is not present throughout the exposure three images are formed, one at each end of the deflection, plus the normal 'undeflected' image, Fig. 16a. As the strength of the deflection decreases the image will be changed from clear fringes to a blurred single fringe, Fig. 16b.

(d) **DC fields** produce test photographs that are made up of two images, the normal image and the deflected image, Fig. 16c. If the magnitude of the pulse from the DC field varies, multiple images may be formed during an exposure.

6 Routine procedures

In the 1960s instruments were cleaned every week – anode, cathode, condenser apertures, objective apertures, fixed apertures, and specimen holders. With more modern instruments, the method of construction combined with a considerable improvement in vacuum pumping fluids has removed the need for weekly maintenance. If we reduce the number of times we open the column for maintenance, we reduce the amount of contamination from cleaning materials and solvents. The vacuum level improves, and the contamination rate decreases. However with any instrument you must be aware of the signs which indicate the need for cleaning.

6.1. The electron gun

Signs which indicate cleaning required:

(a) **High voltage instability**: emission current fluctuating; illumination intensity changing; caustic pattern unstable (Section 1.9); increased time for the emission current to settle after switching on; a high dark current (Section 1.1); inability to reach, or hold, a high kV.

(b) **The filament breaks**: no emission current when the filament control is operated over a range of bias positions.

In case 'a' the gun chamber should be opened; it will almost certainly smell (Section 3.6). The chamber and all gun components should be thoroughly cleaned using the techniques and criteria outlined in Sections 3.7 or 3.8 for the chamber, and Sections 3.1 to 3.5 for the cathode, anode, and corona shield (see Fig. 17 for the names of the gun components).

In case 'b' follow the procedure listed in Section 6.2.

6.2. Changing a filament

(a) Switch off the high voltage and turn down the filament control.

(b) Open the gun chamber to atmospheric pressure.

(c) If no *shorting bar* is provided use a long insulated-handle screwdriver to short out any electrostatic charge, Fig. 18. Touch the metal shank against the chamber edge and angle it across to touch the cathode assembly whilst keeping pressure on the metal chamber edge. Be careful not to scratch the cathode.

(d) TAKE CARE – THE CATHODE MAY BE VERY HOT if a filament has only

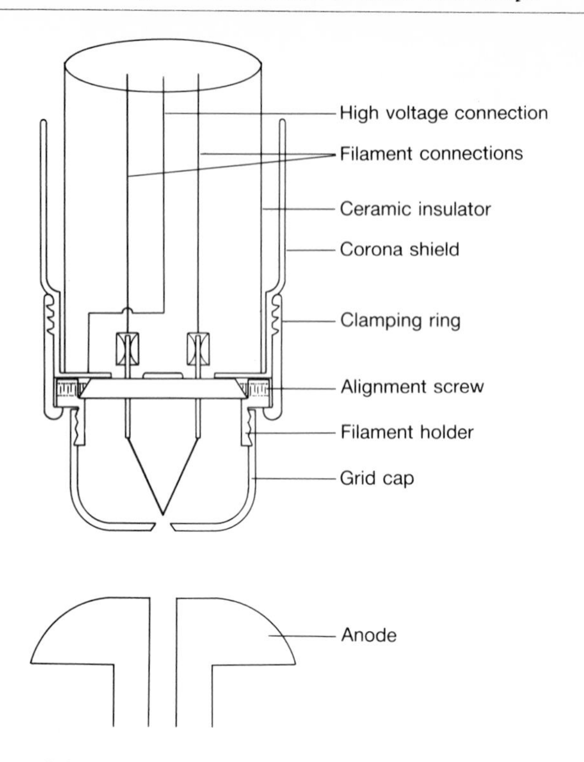

Fig. 17. The components of the electron gun.

just broken. Remove the cathode assembly using a *chamois leather* glove or leather.

(e) Clean the cathode assembly by one of the techniques described in Sections 3.1 to 3.5. Save the old filament for evaluation, see Section 6.7.

(f) Use the information in Sections 3.7 and 3.8 to decide if the chamber should also be cleaned.

(g) Fit a new filament using the manufacturer's recommended setting for the distance between the filament and the grid cap (see also Section 6.6).

(h) Replace the cathode assembly and check the floor of the chamber for debris. Check that the gun O-ring is free from dirt, close the gun chamber and evacuate. Wait for an extra ten minutes after the evacuation is complete to allow the system to outgas before switching on the high voltage.

6.3. High voltage level

A variation in high voltage level will require a variation in lens settings (Chapter 1). When a change of this kind occurs, other set conditions should be checked to see if they too have changed. If the fault (higher or lower lens settings to reach a specific condition) is repeated throughout the checks, then the source is either the high voltage electrical circuit or the common power supply.

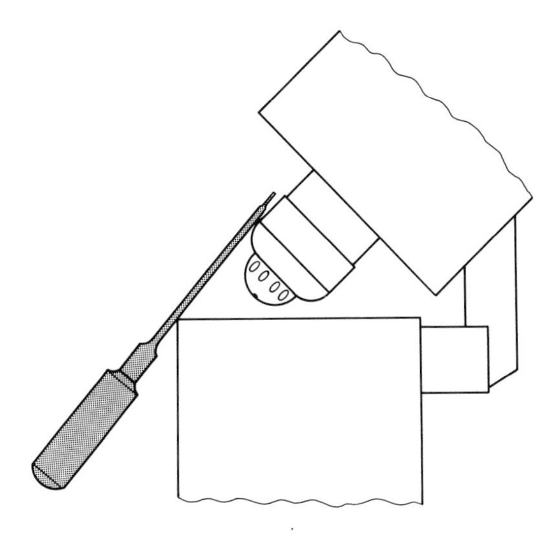

Fig. 18. Earthing the electron gun with a screwdriver.

6.4. High voltage discharge

If high voltage discharge is heard in the gun chamber or in the high voltage tank it may be cured by cleaning (Section 6.1), or by improving the vacuum level by detecting leaks in this area (Section 7.7). Discharge may also mean the failure of a component within the tank and the test described in Section 6.5 may be used to isolate the problem. However discharge in the high voltage tank does not always mean the problem is in the tank itself.

Component failure within the high voltage power supply will often cause discharge within the tank; check the power supply board following the advice given in Sections 8.8 or 8.9. Use the test in Section 6.5, listening for discharge or watching for the corresponding pulse on the emission current meter. It is unlikely that complex tank failures will be corrected without the assistance of the manufacturer's service engineer, so do not try to work in the tank without expert advice.

HIGH VOLTAGE TANKS OFTEN CONTAIN CAPACITORS WHICH WILL HOLD A VERY HIGH CHARGE FOR WEEKS. OPENINGS AND ENTERING A HIGH VOLTAGE TANK IS *NEVER* FREE FROM HAZARDS.

Oil filled high voltage tanks do not require routine maintenance. Should you work with a tank it is most important not to contaminate the oil. At least two people will be required to carefully lift the top plate clear of the tank; spillage is inevitable so surround the area with absorbent paper. Keep the tank covered whilst working on the components, which will in most cases be attached to the top plate of the tank. When lowering the top plate back into position allow time for the oil to expel air from in and around the components. When the tank is fully reassembled rock it slightly to encourage any trapped air to rise to the surface. Run the high voltage at each kV for at least one hour. This will warm the tank components, driving off any air that would act as a conducting medium. Do not rush this *conditioning* procedure or violent discharge and component damage may result.

Freon gas-filled high voltage tanks require routine maintenance. The gas pressure should be checked each month and maintained at the manufacturer's recommended level. When a high voltage discharge has occurred the tank will require *flushing* to remove the polymerized freon. Attach a freon supply to the inlet position which will be near to its base. Open the exhaust valve situated near the top of the tank, and allow the gas to flush through for about 5 min. Close the exhaust valve before increasing the pressure to the manufacturer's recommended level.

If you need to open a freon filled tank the following procedure should be used to repressurize the system:

(a) Attach a rotary pump to the tank and pump to the manufacturer's specification or 0.1 mbar

(b) Replace the rotary pump with a cylinder of freon gas and fill the tank to the desired pressure level.

(c) Evacuate with the rotary pump as in 'a'.

(d) Repressurize as in 'b'.

(e) Flush the tank as described earlier and repressurize to the recommended level.

A normal laboratory 'freon duster' will contain sufficient gas to routinely repressurize a tank. A gas cylinder is ideal for a complete gas change should you need to open the tank. A freon filled tank, if refilled as described above, should not require conditioning for more than a few minutes at each kV.

6.5. Isolating gun problems

Removing the gun cable and gun chamber from the high voltage circuit will help to determine whether the problem is generated within the electronics and the high voltage tank, is due to the gun chamber (dirty or poor vacuum), or due to the high voltage cable itself breaking down (likely in instruments greater than ten years old). If the high voltage is pulsing (Section 1.9), or is heard to be discharging, proceed as follows:

(a) Switch off the high voltage, earth the electron gun, and repump the gun chamber.

(b) Unplug the high voltage cable from the tank, taking great care not to allow dirt to fall into the oil, and cover the empty socket.

(c) Switch the high voltage on at the test kV and watch the emission meter or listen for discharge.

(d) If the meter remains steady or the discharge does not recur, the fault must lie in the gun chamber, gun, or cable. If the fault remains it is the tank or the control circuit (Section 8.10) that is at fault.

If the problem is indicated to be within the chamber or cable, thoroughly clean the chamber and gun components as described in Sections 3.1 to 3.8. Check the gun vacuum level as in Section 7.7. If, after checking and cleaning these areas at least twice. The problem remains, it is almost certainly a fault in the cable unit.

6.6. High voltage problems

Day to day high voltage problems will nearly always occur because of dirty or poor vacuum. Keep the gun chamber clean and leak free by taking great care when changing filaments, or when maintaining this part of the system.

To aid microscopy in general as well as fault finding, it is a good idea to standardize the filament-to-grid cap setting. With a fixed bias condition, if the distance between the filament and the grid cap is varied the emission current will also vary. To correctly set the filament-to-grid cap distance with regard to the manufacturer's recommended emission current, the filament should be moved towards the cap when the current is too low, or away from the cap when the current is too high. This procedure should be carried out with the bias, or emission control, in the middle of its range.

6.7. Filament breaks

Filament breaks are a guide to the way the filament has been used therefore each broken filament should be studied with a hand lens or a microscope. Typical filament breaks are illustrated in Fig. 19.

(a) A **normal break**, the filament thins on both sides breaking on one side with tapered ends.

(b) If the filament has been **overheated** or **oversaturated** it will break as in 'a' but with balls of molten metal forming at the break. Alternatively the molten balls may have fallen off leaving squared ends.

(c) A filament breaking due to high voltage **discharge** has its tip 'blown away'.

(d) When there is a leak in the gun chamber the excess gas causes oxidation of the filament at a very rapid rate; known as **gas attack**. In this case the break looks normal but the filament is considerably thinner and its life will have been very short.

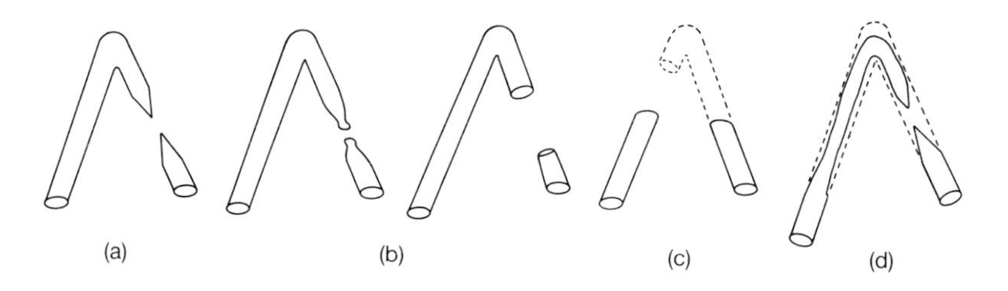

Fig. 19. Types of filament failure: (a) normal, (b) overheated, (c) discharge, (d) gas attack.

6.8. Condenser lens system

(a) **Illumination instability**: illumination drift; charge–discharge conditions (Section 5.8).

(b) **Aperture instability**: illumination drift; charge–discharge conditions. Use another aperture to determine if the problems are due to the lens or the aperture (see also Section 5.8).

Every effort should be made to determine if the fault may be easily cured, for example by changing an aperture, or whether the column requires dismantling. If a blade of the anti-contamination device is situated above the specimen, check its cleanliness. If you encounter a problem whilst viewing a specimen, try another specimen holder, in case it is the upper part of the holder which is dirty and causing the problem. If these checks confirm the dirt to be within the top part of the condenser system the first items to check are the anode and the first fixed condenser aperture. If the 'smaller aperture' test isolates the dirt to the lower part of the condenser lens, this may often be reached by removing a screening tube. If the dirt is not isolated by these methods, the condenser lens system may have to be disassembled, quite a complex procedure. Follow the techniques outlined in Section 6.15, using the instrument instruction book and any cross-section diagrams that may be provided.

6.9. Aperture instability

This may be due to a dirty aperture, which should be replaced, or contamination of the aperture holder itself. Dirty aperture systems will make alignment very difficult due to a beam deflection caused by the dirt charging up. Clean the apertures as recommended in Sections 3.9 to 3.11 and also the holder unit, using an ultrasonic clean if possible.

6.10. Aperture mechanisms

Aperture mechanisms need to be clean and very precise in their action. The O-ring should be cleaned, lightly greased (see Section 7.4), and the mechanism checked for tightness. Different manufacturers use different techniques to make these systems positive in their action. To tighten an aperture unit carefully disassemble the unit, making notes of positions and tensions of any clamps or springs. Wash the components clean with a solvent and reassemble, adjusting the tensioning for a tight smooth action and remember that when under vacuum the unit will be drawn into the microscope by the pressure differential.

6.11. Specimen area

(a) **Specimen instability**: due to a dirty grid or to a dirty specimen holder, see Section 5.2 for evaluation of this area.

(b) **Beam instability**: due to a dirty cold finger or specimen holder; only the areas *above* the specimen will cause *beam* instability.

Specimen holders should be routinely cleaned at least once a month, except where otherwise directed by the manufacturer. If instability is occurring, the time between cleaning sessions is too long. Cleaning is best carried out using an ultrasonic cleaner, see Section 3.5. An anti-contamination device above the specimen will itself cause beam instability with contaminated. An anti-contamination device with an aperture below the specimen will cause gross astigmatism when contaminated. The blade is removed from the main unit for cleaning by one of the methods described in Sections 3.1 to 3.5.

6.12. Specimen exchange system

(a) **Top entry** specimen exchange systems use a cartridge as the specimen holder; this is taken by a transfer device through an airlock and placed into the specimen stage. The transfer device will have both moving components and moving O-ring seals which will require periodic maintenance. Insufficient lubrication on the moving O-rings will make the exchange and transfer process very stiff. When this occurs, the device should be removed from the instrument and all moving seals cleaned and regreased, see Section 7.4. The contact surfaces for the moving seals should be cleaned with ether or acetone. Mechanical movements within the device should not be lubricated in any way, but all of the components should be washed in a solvent to remove contamination.

(b) **Side entry 'cartridge'** specimen exchange system, transfer the specimen by means of an exchange rod. The specimen is mounted within a cartridge which attaches to the rod by means of a small jaw, this is opened and closed by pressing the end of the unit. The rod acts as the transfer device as well as activating the specimen airlock. For stability the cartridge is detached from the rod during specimen observation. The O-ring seal will require periodic cleaning and regreasing and the rod should be kept clean using a solvent. It is most important to open the jaw prior to approaching the specimen cartridge when removing the specimen, this prevents damage to the jaw and the clamping pin on the cartridge.

(c) **Side entry tilt holders.** Most side entry stages that allow the specimen to be tilted have the specimen mounted at one end of a rod; this also acts as one of the stage drive motions. The rod has a location pin which is used to activate an air lock in the column. The O-ring on the rod must be clean and greased for the stage movement in this direction to be smooth and to facilitate specimen exchange. The jewel at the tip of the rod should be kept clean and protected when the holder is withdrawn from the instrument. Only use a soft lint-free cloth to wipe the jewel. The area of the rod between the O-ring and the jewel is within the column vacuum when the rod is in position, and should be kept clean using a solvent to remove contamination.

6.13. Objective lens

Dirt in the objective lens causes astigmatism, as does dirt at any point between the specimen and the lens. This may largely be prevented by making sure that dirt is not picked up on the specimen holder, as it will fall into the lens as the holder is placed into the microscope. Often the lower blade of the anti-contamination device, or the objective lens aperture unit, may catch the debris and should be checked before opening the column to remove the objective pole piece or screening tubes. (See Section 6.15 for advice on removing pole pieces.) Clean all components as described in Sections 3.1 to 3.5 and the apertures as in Sections 3.9 to 3.11. *Do not use harsh metal polish or an ultra sonic cleaner to clean pole pieces* Section 3.2).

See Section 6.9 for general advice on the maintenance of the objective lens aperture system.

6.14. The imaging system

It is quite rare for the remaining parts of the imaging system to become contaminated. Diffraction, intermediate, and projector lens pole pieces are too far from the electron source to become contaminated. Similarly, debris from specimens is unlikely to fall any lower than the objective lens area. However, should contamination be situated near to the beam path in the lower part of the imaging system it will either cause a charge–discharge DC type image shift (Section 5.8), or image distortion. This area of the instrument may be cleaned by removing pole pieces and screening tubes, either through the viewing chamber, or by lifting off the gun–condenser system for access through the objective lens. Follow the guide given in Section 6.15, using the manufacturer's instruction manual and cross-section diagrams as a guide to component position.

6.15. Lens problems

Use the information from Sections 1.3 to 1.10 to isolate the problem lens; try switching off lenses to see when, and if, the fault is removed. If you suspect that a lens is not operating, check if each magnification step does increase the magnification and refer to information similar to Appendix 1 to check the system. If there is a pulse in the image does it stop when a lens or deflection coil is switched off? If this occurs the problem is within the electronics and Section 8.10 should be followed. If, using Sections 6.8 to 6.14, you are able to trace a fault to dirt in the column but are unable to reach the critical area easily, you must decide if you have the skill to open the column, or require professional help. Remember, electron microscopes are constructed by bolting or fitting components together; they may therefore be disassembled piece by piece, but certain rules must be observed.

(a) Do not try to remove a lens pole piece with the lens switched on.

(b) Do not pull a pole piece or a screening tube out of a lens without removing any variable aperturs or cold fingers that may protrude into the column.

(c) Do not rotate a pole piece as you remove it as it may be marked in some way to ensure optimum positioning.

(d) Take care not to mix pole pieces and their fixed apertures; they may look the same, but are designated for a particular role.

(e) Take care not to damage pole pieces. They are the heart of each lens and damage means image or beam distortion.

(f) Work on each lens unit in turn if possible, removing, cleaning, and replacing, prior to moving to the next area. Work so that clean areas are not made dirty again; cover each area and insert variable apertures to catch debris.

(g) When working on mechanical systems, shifts, aperture units, stage mechanisms, etc, be aware that they will almost certainly contain springs which may fly out as you are dismantling the unit.

(h) Before moving a lens unit as a whole, disconnect vacuum pipes, electrical supply lines, and cooling water. Drain the water lines to prevent water falling onto components.

6.16. Safety

Electron microscope servicing may create hazards, so always have the column checked for X-ray leakage when the gun or condenser areas have been dismantled.

When working with the panels off the instrument exposing the electrical circuits, never work without assistance being within shouting distance. Do not let the manufacturer's service engineer work alone unless similar circumstances prevail.

6.17. The camera and vacuum system

The camera and the vacuum systems which use electric motors are controlled by micro switches. These switches stop the mechanisms at set points within a sequence. If they become loose they will alter the sequence causing problems. In order to adjust this type of system correctly a complete understanding of the mechanism is required. *Do not attempt any re-adjustment until you are absolutely sure of your actions.*

Compressed air driven vacuum systems rely upon an electrical signal activating a solenoid to allow the air to pass to one or other side of the valve. Problems occur either through the solenoid failing to operate, or the diaphragms within the valve sticking or leaking. Leaking valves may be repaired using a kit obtainable from the manufacturer; sticking valves are often due to a damaged seal.

6.18. Fluorescent screens

These will deteriorate with use and in general it is advisable to change the screen, or screens, each year. As a screen ages, its intensity falls making positions of exact focus and astigmatism more difficult to determine.

6.19. Viewing windows

Viewing windows should be cleaned externally with household cleaner, ether, or acetone. Do not clean the inside of windows with anything other than ether or acetone, as they are metal coated to prevent charging. Check for the coated face by using an ohm meter: the low resistance side is the conducting surface.

7 The vacuum system

The transmission electron microscope's vacuum system has become more important as the quest for higher resolution, and a cleaner specimen environment, has accelerated. The trend toward running the vacuum system continuously has placed more emphasis on its reliability, and therefore routine maintenance plays a more critical role.

7.1. The rotary pump

The rotary pump is the first unit in the vacuum system. Maintenance of the pump is minimal but it is essential to check the oil level, and belt tension if appropriate, each month. Other points to watch are oil leaks, the most likely source on a belt-driven pump being the oil seal behind the drive pulley. To replace the seal:

(a) Block the air inlet and pumping line, then lay the pump on its back.

(b) Remove the pulley and hook out the seal with a screwdriver taking care not to scratch the critical shaft face.

(c) Grease the new seal inside and out, then using three or four sharpened orange sticks to guide the seal over the shaft lip, ease the seal into position.

(d) Tap home the seal using a block of wood and a hammer.

(e) Replace the pulley firmly, top up with oil, and reconnect the pump to the vacuum system.

The fluid in the rotary pump will require changing every year if the pump is used continuously. As the fluid provides a vacuum seal, lubrication, and corrosion protection, frequent changes will prolong pump life. The fluid may be purged with air to remove condensed vapours by isolating the pump from the system and opening the **gas ballast** valve. Pumps with this facility should be ballasted for about fifteen minutes every month. Always vent the pump to outside the laboratory or fit an oil mist trap. The manufacturer's information will advise on ballasting, fluid type, and quantity.

7.2. The air compressor

The air provides the activation force to operate vacuum valves in most modern instruments. The oil level should be topped up to the mark and the air reservoir drained of condensed water each month. The oil should be changed annually. Moisture is the greatest problem in these systems as it collects in the activation areas of valve control units causing corrosion and leaks.

7.3. Diffusion pumps

These are fairly trouble free and the only problems that may occur are corrosion due to condensation from over cooling, or a heater failure. In order to reduce corrosion, the water temperature should be held within ± 2°C of room temperature, at a flow rate that allows the second water cooling coil from the base of the pump to feel slightly warm to the touch. The diffusion pump heater may be checked as described in Section 7.9.

7.4. O-rings

O-rings found at every joint in the vacuum system and they should only be removed with a wooden stick to prevent damage to the ring or seat. Fixed position O-rings should be cleaned with a solvent or hot soapy water. Prior to replacing an O-ring, run it between your thumb and finger placing the ring under slight tension. In this way minor defects, cracks due to radiation damage, or tears due to damage on assembly will be made more clear. Clean the O-ring seat with solvent and replace the seal, but do not use grease. Moving seals should be cleaned and checked in the same way as fixed position seals, but in this case a very small quantity of grease should be massaged into the seal. The excess grease should be wiped away with Velin tissue. *Do not over grease as this is a source of contamination.*

Seals within pumping control valves will require periodic maintenance, particularly motor driven 'piston' units where the O-rings need liberal greasing, preventing them from binding excessively and ensuring efficient operation. O-ring seats should always be cleaned with a solvent and checked for damage. *Never use a metal tool to remove an O-ring as this will scratch the seat.* A damaged seat may be restored using a medium grade emery paper rubbing *round* in the O-ring direction.

7.5. Vacuum leaks

Vacuum leaks may be first recognized from the vacuum gauge readings, an increase in contamination rate, a fall in filament life, or slower pump-down times. The electron column may be divided into two types of vacuum area.

(a) **Fast pumping areas** (the gun chamber, the specimen area, and the viewing chamber) are adjacent to wide bore pumping lines and are rapidly pumped by the diffusion pump.

(b) **Slow pumping areas** (the condenser lenses, the imaging lenses, and the camera chamber in instruments without a direct diffusion pump line) are separated from the wide bore pumping lines by a narrow bore line or apertures.

Leaks in different areas produce different vacuum gauge readings. For example; a leak in a slow pumping area results in a poor reading on the gauge that monitors the column, but a good reading on the backing line gauge. A leak in a fast pumping area results in an average reading on the column gauge, but a poor reading on the backing line gauge.

Normally leaks only occur suddenly as a result of some action. When a leak has occurred always check the last action that you made. Moving O-rings in aperture mechanisms or the specimen exchange are often the source of a 'sudden' leak.

7.6. Leak detection

After the simple checks outlined in Sections 1.14 and 7.5 have been performed, sections of the column are isolated until the leak is no longer evident. Metal **blanking plates** are used to seal the column as the upper units are removed. At each stage the vacuum system is pumped to a specific level, e.g. 5×10^{-4} mbar, or a point where the 'HV Ready: light is illuminated. Only when the area that contained the leak has been excluded from the system will the required vacuum level be reached. Large diameter rubber bungs are useful to seal pumping lines.

When rebuilding a system clean each seat and seal (Section 7.4), and only regrease moving seals. When bolting units together use the vacuum to pull the system square, prior to tightening the clamping units. *Do not overtighten clamps or bolts: this will not cure leaks.*

7.7. Poor filament life

This is often caused by a vacuum leak in the gun area (see also Section 6.7). Vacuum gauges are normally placed in the main pumping manifold but a good vacuum at this level doe snot always mean that the gun vacuum is satisfactory. The only method for checking the vacuum level in the gun area is to remove the gun assembly and place a blanking plate fitted with a vacuum gauge in its place. Expect a vacuum level better than 10^{-4} mbar within one hour, without the use of liquid nitrogen traps.

7.8. Poor diffusion pump performance

A pump that becomes sluggish in operation but has no vacuum leak probably requires cleaning. If the pump has not been cleaned for two or three years this could also be the reason for a slower pumping speed although it may achieve the same ultimate vacuum. To clean the pump follow the procedure listed below:

(a) Let air into the column and the *cold* diffusion pump and switch off the instrument. It is not always possible to get the instrument to allow air into the diffusion pump. You may be forced to let air into the pump by easing off the pumping line which runs between the diffusion pump and the rotary pump. This may mean opening the connection between the backing tank and the diffusion pump.

(b) Remove the panels in order to obtain access to the pump.

(c) Remove the rotary pump or backing tank vacuum line.

(d) Remove and drain the water lines.

(e) Remove the heater connections and unbolt the pump; take care as it is quite heavy and you will probably be handling it at a difficult angle. It may be possible to take the weight of the pump by packing blocks of wood between the pump and the floor.

(f) Empty the pump and measure the contents to evaluate fluid loss; up to 25% of the figure indicated is normal.

(g) Remove the jets and clean them with metal polish; difficult stains may be removed with emery paper. Take care not to round the edges of the vanes. Wash all the pieces in hot soapy water. Clean the O-rings and seats as in Section 7.4.

(h) Refit the vanes and refill the pump with the correct amount of fluid. Warm the measuring cylinder containing the fluid in a bath of hot water or with a hot air blower, to ensure that all the fluid flows into the pump. Do not overfill as this decreases the pump efficiency, and use the correct fluid as some old heater wattages may not be sufficient for modern high temperature fluids.

(i) Refit the pump, loosely tightening the bolts; refit the rotary pump or backing tank connection; connect the water lines (check for leaks); connect the heater and evacuate the diffusion pump. TAKE GREAT CARE NOW THAT THE MAINS ARE ON. DO NOT WORK UNLESS HELP IS WITHIN SHOUTING DISTANCE. Tighten the bolts as the vacuum seats the pump onto its flange.

(j) Set the normal vacuum sequence in motion and check that it is operating satisfactorily.

(k) Switch the instrument *off* overnight. Then check for leaks in the diffusion pump area by listening for sounds of the rotary pump pumping air when it is first switched over to evacuate the diffusion pump. If the rotary pump does pump air the diffusion pump should be removed, all O-rings checked, seats cleaned, and then procedures 'i' to 'k' followed.

7.9. Diffusion pump heater failure

Diffusion pump heater failure often causes concern; a series of checks may be made with a multimeter if you suspect this problem.

(a) Remove the panels to obtain access to the diffusion pump.

(b) TAKE CARE! Check the voltage applied to the diffusion pump using the meter set to AC volts (see Section 8.7). Place a circuit probe on each of the two heater contacts. Japanese instruments will read 100 volts, European instruments 220 volts, UK-built instruments 240 volts.

(c) If the voltage is zero, check the vacuum process to see if the pump should be in a 'heating' condition. If the pump should be operating check its fuse. Check procedure 'd', below, prior to fitting a new fuse.

(d) If there is a correct voltage on the heater but the pump is cold, switch off the diffusion pump or close down the instrument. Remove the diffusion pump heater connection in order to check the heater resistance. Set your multimeter on resistance and place the connections onto the heater terminals, it should measure 20–100 ohms. If this reading indicates an open circuit, a short circuit, or a connection between one of the terminals and the pump casing, the heater element has failed.

(e) Most elements are either bolted to the base of the pump or sit in a retainer which holds the element against the pump base. Heaters become warm very quickly: take care when working near the diffusion pump.

8 Electrical and electronic maintenance

The electrical and electronic circuit in a transmission electron microscope is basically as shown in the fold-out section at the back of the handbook. The circuit is made up of a number of units which provide and control specific voltages in order to power the systems within the instrument.

8.1. The input and step down unit

The input and step down unit is supplied from the laboratory mains. Instruments manufactured in the UK or Europe operate directly from 220 to 240 volt mains. Japanese instruments operate through a step down transformer using a basic supply of 100 volts. Some instruments have a mains stabilizer built into the input to the microscope in order to reduce the variations in mains voltage.

8.2. Common power supply units

These take the supply from the input and step down unit and generate higher or lower voltage outputs which are used to drive the control circuits.

Prior to the mid 1960s valves were used in control circuits, and common supply outputs of up to 800 volts were required. Between the mid 1960s and the early 1970s transistors gradually replaced valves and the common supply ouputs were reduced to about 100 volts. With modern solid state devices, common supplies develop outputs of less than 70 volts.

Both common power supply units and control circuits use a feedback system. The system compares a voltage feedback from the unit being controlled with a reference voltage, adjusting the level to retain a constant output. Reference voltages were originally provided by batteries but most instruments produced after the early 1970s generate their own reference voltages, often using a temperature controlled reference unit.

8.3. Solid state devices

All modern electron optical instruments now use solid state devices. They are:

(a) **Integrated circuits** (IC): semi-conductors that may contain a number of transistors, a single circuit, or a number of similar circuits. These devices may look like a round transistor but have up to twelve legs. More modern IC's are packaged in black boxes with up to sixteen legs. IC's are broken down into two major categories:

Analogue IC: used to amplify, process, or operate on input signals.

Digital IC: used in logic and computer systems, some of which are known as microprocessors.

(b) **Microprocessors**: semi-conductors that contain memory circuits; PROMs – Programmable Read Only Memory. RAMs – Random Access Memory. They are much larger than analogue IC's with up to sixty legs. Both types of device are liable to *damage* if removed from a circuit. A PROM may be damaged by static electricity and a RAM will lose its memory as soon as the supply voltage is removed.

PROMs are programmed by the instrument manufacturer, a typical application being to select the appropriate lens combinations in order to achieve a distortion free image throughout the magnification range.

RAMs are used to memorize user information, e.g. specific lens combinations or specimen stage position. A rechargeable battery may be installed within the RAM circuit in order to retain information even when the instrument is switched off.

8.4. Control circuits

Control circuits supply the power to the column units responding to signals from the desk controls. Circuits containing valves use high voltages to generate low currents, typically 20–750 mA for a lens circuit. Solid state devices use very low voltages generating lens currents ranging from 1–6 A.

8.5. Circuit layout

Circuit layout varies depending upon the age of the microscope. The old valve systems produce a large amount of heat and as the components themselves are very large, valve circuits are very bulky. In order to dissipate the heat the circuits are either mounted horizontally in drawers or vertically, bolted directly to the power console. Airflow is important as cooling is achieved through convection or by means of fans. Some manufacturers prevent access to 'live' circuits by fitting safety switches which remove the power from the circuits when panels are opened. Modern solid state circuits are very compact allowing considerable variation in presentation by the manufacturers.

Common power supplies will all be mounted together, even if they are switched separately to the vacuum control circuits. Lens and beam deflector supplies may be small separate plug-in boards or mounted on a single large board. In the latter case it is common to group all the lens circuits on a single board. The beam deflection and stigmator circuits may be similarly grouped. In some cases circuit boards may be mounted away from the main circuit groups, e.g. the high voltage selection board or the film numbering circuit.

When plug-in boards are mounted side by side, an extension board is required if adjustments or checks are to be made. These boards have a plug at one end and a socket at the other. The circuit in question is removed from its socket and replaced by the extension board. The circuit board is then plugged into the extension board where it will operate normally whilst allowing access.

8.6. Monitoring the electronics

WORKING WITH THE ELECTRICAL SIDE OF AN ELECTRON MICROSCOPE IS VERY DANGEROUS. DO NOT INVESTIGATE THIS AREA UNLESS YOU ARE CONFIDENT THAT YOU WILL BE ABLE TO WORK SAFELY. WHEN WORKING ANYWHERE ON THE INSTRUMENT THAT REQUIRES REMOVAL OF A PANEL, TAKE GREAT CARE NOT TO TOUCH OPEN CONTACTS. REMOVE WATCHES AND ANY OTHER METAL ITEMS FROM THE HANDS AND WRISTS. NEVER WORK ALONE. ENSURE HELP IS WITHIN SHOUTING DISTANCE THROUGHOUT YOUR INVESTIGATIONS.

The supply entering and leaving each unit within the electronics may be monitored and used as a guide when fault finding. The ideal test instrument is an electronic multimeter, which draws a very low current and will not load the very sensitive circuits now in use. AVO or moving coil type meters should only be used for mains voltage and continuity checks. An oscilloscope able to measure down to one millivolt is a useful tool for tracing instabilities within a circuit.

8.7. Using the test equipment

Prior to taking any voltage reading, always position the leads in the appropriate sockets (AC or DC), and set the meter on the highest voltage range. Reduce the meter range when you are satisfied that the measured value will not exceed full scale.

When measuring the voltage in an AC circuit the two connections, live and neutral, may be measured relative to earth (one probe attached to the earth terminal or chassis) or relative to each other.

When measuring DC circuits there is only a 'live' voltage which is related to a **zero volts** connection. The circuit boards on most instruments have the '0 volts' or 'earth' clearly marked. In some cases a group of test points will use a specific zero volts connection e.g. + 15 V and – 15 V may share a 0 volts test point mounted between them on the circuit board.

Using an oscilloscope the **ripple** (peak to peak value, Fig. 20) and **stability** (the change in vertical position of a trace, Fig. 20) may be measured. When taking readings the flying lead attached to the probe should be connected to the appropriate zero volts test point.

8.8. Mains transformer/stabilizer

The mains transformer/stabilizer should have an input within 10% of the normal mains 240 volts. Test readings between live and earth should read approximately

Fig. 20. Analysis of oscilloscope traces.
Ripple – the voltage variation displayed by an oscilloscope.
Stability – the consistency of the trace position.

240 V, with less than 2 V between neutral and earth. If the latter is high, arrange to have the *live and neutral* connections reversed; a high neutral may cause instability. The supply will be fitted with a fuse which will fail should the transformer become overloaded.

The output from the unit should be within ± 5% of the appropriate voltage: 240 volts for UK equipment, 220 volts for European equipment, or 100 volts for Japanese equipment. Each instrument will have a means of adjusting the output, either by a variable control or through movement of the connections on the transformer to a higher or lower voltage setting.

8.9. Common power supply units

The AC input to the common supply units is set by the manufacturer as listed in Section 8.8. Fuses may be placed on the input or the output of each supply module. Each output will be a DC voltage, the following values being typical of a modern power supply unit:

+ 5 volts. . . . . within 0.2 V (4%)	maximum ripple 2 mV;
+ 15 or − 15 volts . . . within 0.5 V (4%)	maximum ripple 2 mV
+ 24 volts. . . . . within 0.5 V (2%)	maximum ripple 2 mV
+ 70 volts. . . . . within 2 V (2.8%)	maximum ripple 2 mV

In each case the stability should have no visible variation.

A large change in common supply voltage will be due to a component failure within the particular supply circuit. This may also cause a fuse on the input to the circuit to break, or in some cases it may damage components in the control circuits.

8.10. Control circuits

Control circuits have a range of inputs: from the common power supply, a signal from the desk controls, a feedback from the unit, and an output to the unit's power transistors. Typical output values for modern control circuits are:

Alignment (Deflection Coils): − 6 to + 6 V depending upon the control position.

Lens output (Objective at focus): + 1.2 to + 1.8 V depending upon the high voltage and magnification.

Lens output (First projector): + 0.5 to + 1.5 V depending upon the high voltage and magnification.

High voltage (Maximum kV): + 68 V, maximum ripple 0.8 V with no visible variation in stability.

Stigmator + 5 to − 5 V depending upon control position.

8.11. Tracing faults

When your investigations lead you to suspect the electronics, sensible procedures should help you to narrow the problem down still further. It is not too difficult within the scope of this chapter to outline procedures relating to a circuit failure. Unfortunately the most common electronics problems, instability and partial circuit failure, require a much deeper electronics knowledge than may be conveyed here.

Figure 21 is a simplified diagram of a typical lens circuit. The current flowing

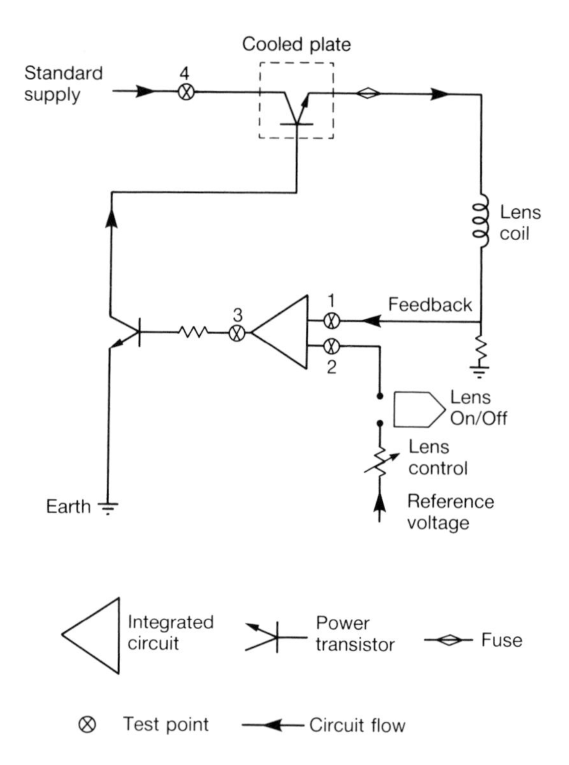

Fig. 21. A typical lens supply power circuit. Test points 1, 2, and 3 should vary with a change in lens current. Test point 4 will remain constant.

through the lens coil is measured across the fixed resistor. The voltage across the resistor is compared with a reference voltage the value of which is determined by the accelerating voltage being used. The difference between the two voltages is amplified and used to control the current flowing through the lens coil by regulating the current output from the power transistor(s). The effect is a balancing of the voltage across the fixed resistor with the applied reference voltage. Trouble shooting flow chart A (Appendix 3) demonstrates fault finding procedures suitable for this circuit.

Figure 22 is a diagram of a typical beam deflector or stigmator circuit. Standard supplies are fed to the circuit along with the reference voltage which is adjusted by means of the desk controls. Trouble shooting flow chart B may be used to fault find in this type of circuit.

The high voltage circuit is very complex; fault finding may be assisted by using Trouble shooting flow chart C, but electronics checks may be limited to comparing test point readings with a set of standard values.

In each of the cases mentioned the circuit boards will almost certainly be marked to indicate the standard voltages e.g. + 5, + 15, − 15, + 24, 0 volts. Test points in lens or deflection coil circuits may be deduced from the circuit diagram. Relate the position of the coils to the circuits shown in Figs. 21 and 22 working back to the output and hence its test point.

When tracing manufacturer's circuit diagrams be aware of some of the techniques used.

(a) Common links, Fig. 23, are often used to take a group of connections from one point in a circuit to another.

(b) Connections to a plug or socket may be coded in relation to the plug or socket, e.g. the connection to socket J4 pin 11 may be coded 411.

(c) The number on a plug or socket may not be the number used within the circuit diagram e.g. the socket for plug numbered A234 may be socket number S1 on that particular circuit diagram.

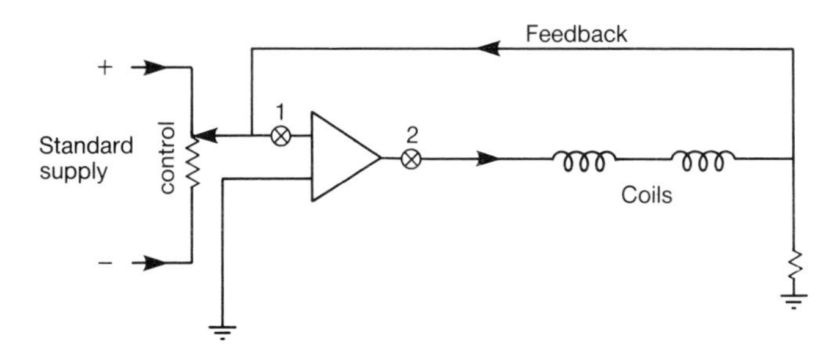

Fig. 22. A typical beam deflector or stigmator circuit. Test points 1 and 2 should vary with a change in control position.

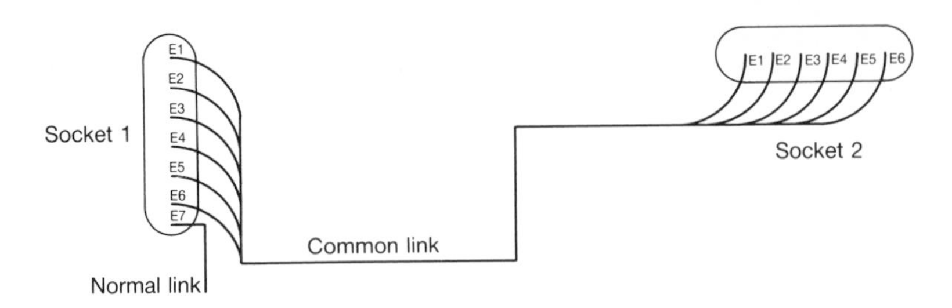

Fig. 23. An example of a 'common link', indicating a multiple connection by a single line on the circuit diagram.

8.12. Valve systems

Faults may be found very easily in a valve system. A flashing valve will produce a 'flashing' image. A valve that fails to light will cause a problem in that circuit. In a case of instability, have an assistant tap the valves within a suspect circuit with an insulated screwdriver. Either watch the image to detect an increase in instability, or observe an oscilloscope trace of the circuit output. Faulty valves have very little resistance to shock therefore this test will exaggerate a fault. Once a fault is traced to a particular circuit it is often simplest to replace all the valves if a valve testing instrument is not available. Substituting the valves one by one will not find the fault if several valves are working incorrectly. But replacing the old valves one by one will isolate the faulty units.

8.13. Transistors

Transistors within circuits are rarely traced to be the source of faults without considerable electronics servicing experience. However power transistors on the output of a circuit, or mounted on a cooled plate, are the most likely fault when a circuit fails. Take great care if transistors are soldered into a circuit. If you are capable of removing the suspect transistor carry out the test indicated in Fig. 24. Set your multimeter on ohms. Connect between any two pins and noting the reading reverse the connections. A 'good' transistor will have two pairs of readings of low resistance combined with reverse connections that measure a very high resistance. The third pair of readings will both be very high resistance. If the transistors are marked you will see that the readings correspond to Base–Collector and Base–Emitter with the double high resistance readings being Emitter–Collector. Damaged transistors will not produce these readings, but if in doubt change the component.

8.14. Integrated circuits

Integrated circuits are most simply evaluated by replacement. If a circuit is found to be faulty only consider changing the IC's if they are plugged in. All the IC's in

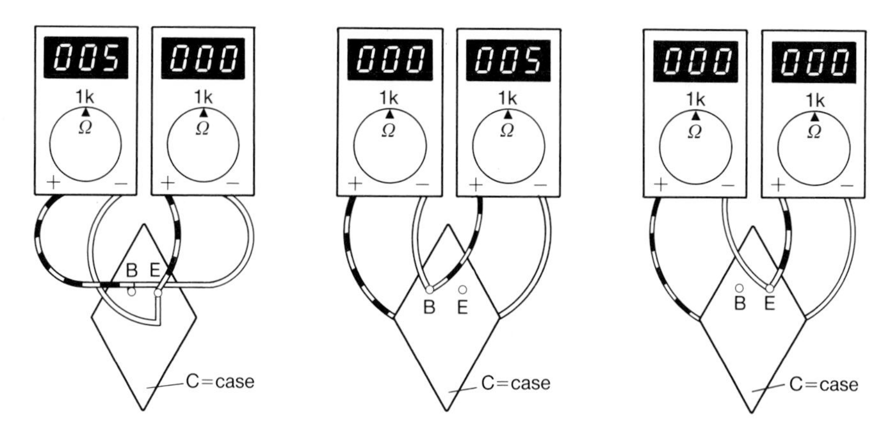

Fig. 24. Procedures for testing power transistors.

the circuit may be changed, or replaced one by one until the fault is removed. Special tools are available for removing an IC, these are recommended as the contacts are easily damaged.

8.15. Replacement components

Fit only **identical** or **direct replacement** components. Never fit a fuse of a higher value in any circuit.

> ALWAYS SWITCH OFF THE ELECTRONICS WHEN REPLACING COMPONENTS OR REMOVING CIRCUIT BOARDS. DOUBLE CHECK YOUR ACTIONS PRIOR TO SWITCHING THE POWER BACK ON.

9 High resolution performance: using the optical diffractometer

The procedures recommended in the previous chapters should enable you to maintain the performance of the TEM at the values guaranteed by the manufacturer, or perhaps at the level demanded by the majority of the users of your instrument. Remember that the vast majority of electron micrographs are aimed at a resolution not better than 2 nm, which is well within the scope of normal checking techniques in which the printed micrograph or, better still, the original negative is examined with a ten times hand magnifier. However, many modern instruments can do considerably better than this and point-to-point resolutions down to 0.2 nm are now available commercially. In the region of 2–0.2 nm of resolution, more sophisticated methods of image evaluation, fault-finding, and routine maintenance are generally necessary.

9.1. Image contrast

The image contrast at the region of 2–0.2 nm of resolution arises mainly from phase contrast effects, i.e. the phase rather than the amplitude of the wave leaving the specimen differs from that of the incident wave on the specimen. This means that the contrast is sensitive to the exact position of focus and to the effects of astigmatism and misalignment of the electron optical column. Although the tests previously mentioned will usually indicate the presence of faults at this level, they are generally not sufficiently conclusive to enable a true diagnosis of the problem to be made. This can be readily understood by very simple considerations. Consider a photographic film of size 100 mm × 80 mm. An individual picture element (pixel) would be of the order of 30 μm × 30 μm. There are thus some ten million pixels in a single exposure. If each of these is to be inspected in turn through a 10× magnifier it would take approximately four months to complete the visual task, apart from the enormous difficulty of recording the information. Even a large computer would have difficulty in handling the data from a through focal series of five such micrographs. What we need is a simple device that would examine all pixels simultaneously and then record all the spacings (spatial frequencies) that are present in the image. The spatial frequency is simply the reciprocal of the actual spacing in the image. Thus the 0.2 nm spacing in the gold lattice has a spatial frequency $0.2\ nm^{-1}$. Now a diffraction pattern has precisely this property, namely that the finer the spacings in the specimen, the larger is the radius at which the corresponding diffraction points occur. Thus an optical diffraction pattern of an electron micrograph will reveal in a convenient and simple manner all the spacings that are present in the micrograph. For example, if there are no spacings present smaller than 1 nm, the optical diffraction pattern will extend from the centre no further than to a radius corresponding to 1 nm^{-1}. Consequently, if we use a specimen containing all relevant spatial frequencies, optical diffractometry gives a rapid, quantitative and unambiguous check on the resolution of an electron microscope. In addition, as will be shown below, the optical diffraction pattern frequently gives a clue as to the cause of any loss of resolution. When used for the analysis of electron micrographs, a diffractometer of simple and cheap construction may be employed, using standard components. An optical diffractometer can also be used, in principle, for subsequent image processing of a micrograph. However there are many practical difficulties in image reconstruction work of this type. It is emphasized that the methods of the present chapter are restricted to micrograph analysis for the *sole purpose* of evaluating microscope and operator performance and of diagnosing the nature of any malfunctioning of the instrument. Image processing itself is better carried out on a digital computer, a topic beyond the scope of this handbook.

9.2. A simple optical diffractometer

Figure 25 shows a simple optical diffractometer that makes use of standard optical components. The source of light is a helium-neon (red) laser of 5 mW output which is bright enough for normal micrographs and yet weak enough that accidental reflections in the eye, should they occur, are not seriously injurious. Naturally the siting of such a laser should follow the normal safety rules for the use of low power lasers. Lens 1 is a standard 16 mm microscope objective lens whose function is to converge the laser beam into a focus at the focal point of this lens, from which the beam spreads out in a diverging cone. An old TEM aperture mechanism with an aperture A of about 50 μm diameter is mounted with axial and X–Y adjustments for accurate centration about the focal point. The function of this aperture is to exclude any unwanted subsidiary beams from the laser which might confuse the optical diffraction pattern. Lens 2 is a 250 mm biconvex lens of moderate quality positioned so as to focus the aperture A on to the screen S by converging the beam. The micrograph for analysis is placed in this converging beam and diffracted beams are produced by the differently spaced details in the micrograph. In the equipment shown, the illuminated patch on the micrograph, and hence the area analysed, will be about 10 mm in diameter (corresponding to about 10^5 pixels). If a larger patch is required, an 8 mm microscope objective lens may be substituted for Lens 1. For the purposes of fault analysis, the more expensive double lens often adopted for Lens 2 in commercially available equipment to ensure parallel incidence at the specimen, is quite unnecessary. In addition, with a single lens, the convergence of the incident beam allows the micrograph to be moved axially, thereby altering the size of its diffraction pattern to a convenient

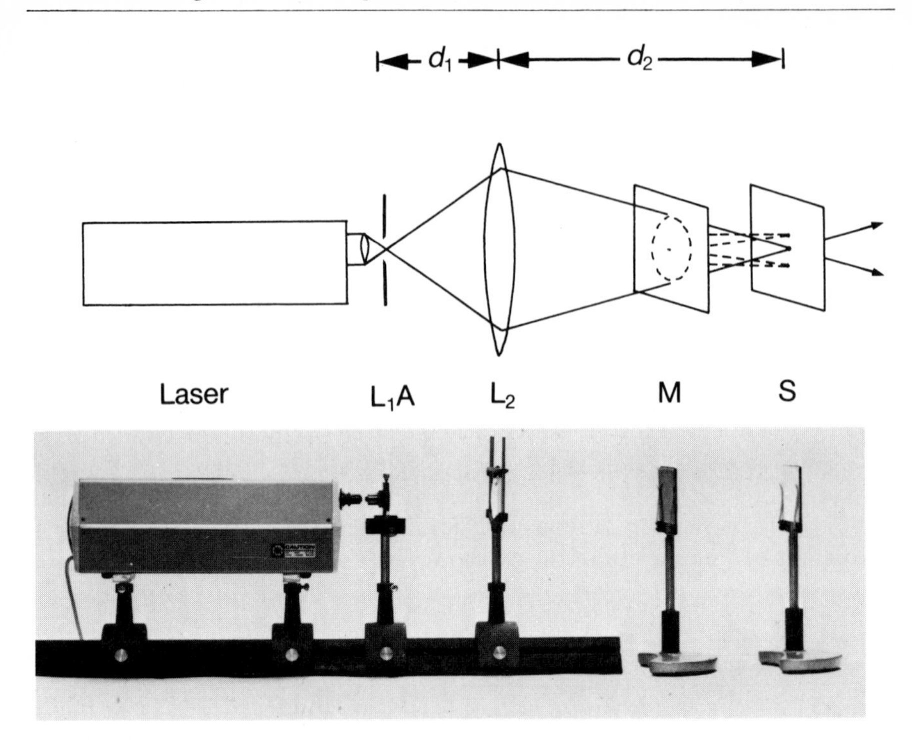

Fig. 25. A simple laser diffractometer made from standard optical components.
d_1 = distance between focal plane aperture A and main lens L_2
d_2 = distance between main lens L_2 and screen S
L_1 = microscope objective (auxiliary converging lens)
M = specimen micrograph

value, e.g. for photography, without the need for any refocusing. The diffraction pattern (dashed rays) appears on the screen S. An important practical point is to provide a pinhole in the centre of the screen S so that the bright central spot passes through it thereby improving the contrast of the diffraction pattern by eliminating the glare caused by the bright central spot. Photography is straightforward; the screen is replaced with a camera-back (i.e. a camera with the objective lens removed) and the diffraction pattern is photographed directly onto the emulsion. The distances d_1 and d_2 would be roughly 300 mm and 150 mm for normal use, so a total bench space of 2 m or even 3 m is desirable if a wide range of micrographs of different spacings has to be examined. The optical bench itself, however, can be quite short, perhaps 700–800 mm in length, since only the laser and the two lenses must be mounted rigidly together.

The diffractometer can be calibrated by replacing the micrograph with a diffraction grating of known spacing and photographing the resulting pattern, or simply measuring it on the screen with a ruler. Suppose this grating spacing is 0.025 mm/line, and is positioned between L_2 and S (Fig. 25) so that its innermost (first order) diffraction spots are each 25 mm from the centre of the pattern. Replace the grating with a micrograph of known magnification M, giving a ring diffraction pattern as in Fig. 26, or spots as in Fig. 30. Any feature a distance R mm from the centre will correspond to the distance D mm in the original specimen given by the simple relation:

$$D = \frac{0.025 \times 25}{R \times M} \text{ mm} = \frac{0.025 \times 25}{R \times M} \times 10^6 \text{ nm.} \qquad (9.1)$$

If R is the outer limit of a circular carbon diffraction pattern, then D is the resolution limit of the microscope. If R is the radius of a dark ring, the corresponding D is a spacing not being imaged by the microscope (see Section 9.3). If R is the radial distance of the inner diffraction spot from the image of a crystal (e.g. catalase or a grating replica) for which D is known, then the magnification of the micrograph may be calibrated by rearranging the above equation to give

$$M = \frac{0.025 \times 25}{R \times D},$$

provided all the measurements are in millimetres.

Because of the compact size of this diffractometer it is not necessary for it to occupy bench space permanently as it is easy to set up. Ideally the diffractometer should be installed in a room that can be darkened. The normal safety precautions for low power lasers should be observed. For example the laser should have a key-operated switch and all users should have some preliminary instruction before being allowed to analyse micrographs. A warning notice should be displayed to alert and deter casual visitors. The total cost of such a diffractometer is chiefly that of the low power laser and represents less than one per cent of the cost of the electron microscope.

9.3. Testing the microscope

It is fortunate that the ideal specimen for the critical testing of a high resolution microscope is a normal carbon support film. It has been shown that the typical granular appearance of such a film, caused by phase contrast effects, represents a broad range of spatial frequencies. The range of frequencies actually transmitted by a TEM depends, among other things, on the focal setting of the objective lens, the coherence (angular aperture) of the illuminating beam, the stability of the electronic supplies, the presence of astigmatism, of vibration or specimen drift, and of stray AC magnetic fields. Hence all the factors which limit the performance of a TEM will also restrict the spatial frequency range which is observed on the optical diffraction pattern. Thus by taking a focal series of a carbon support film, with the objective lens focused above and below the normal focusing position, (Section 5.3), a corresponding set of diffraction patterns will be obtained in the optical diffractometer. These will reveal the capability or otherwise of the TEM

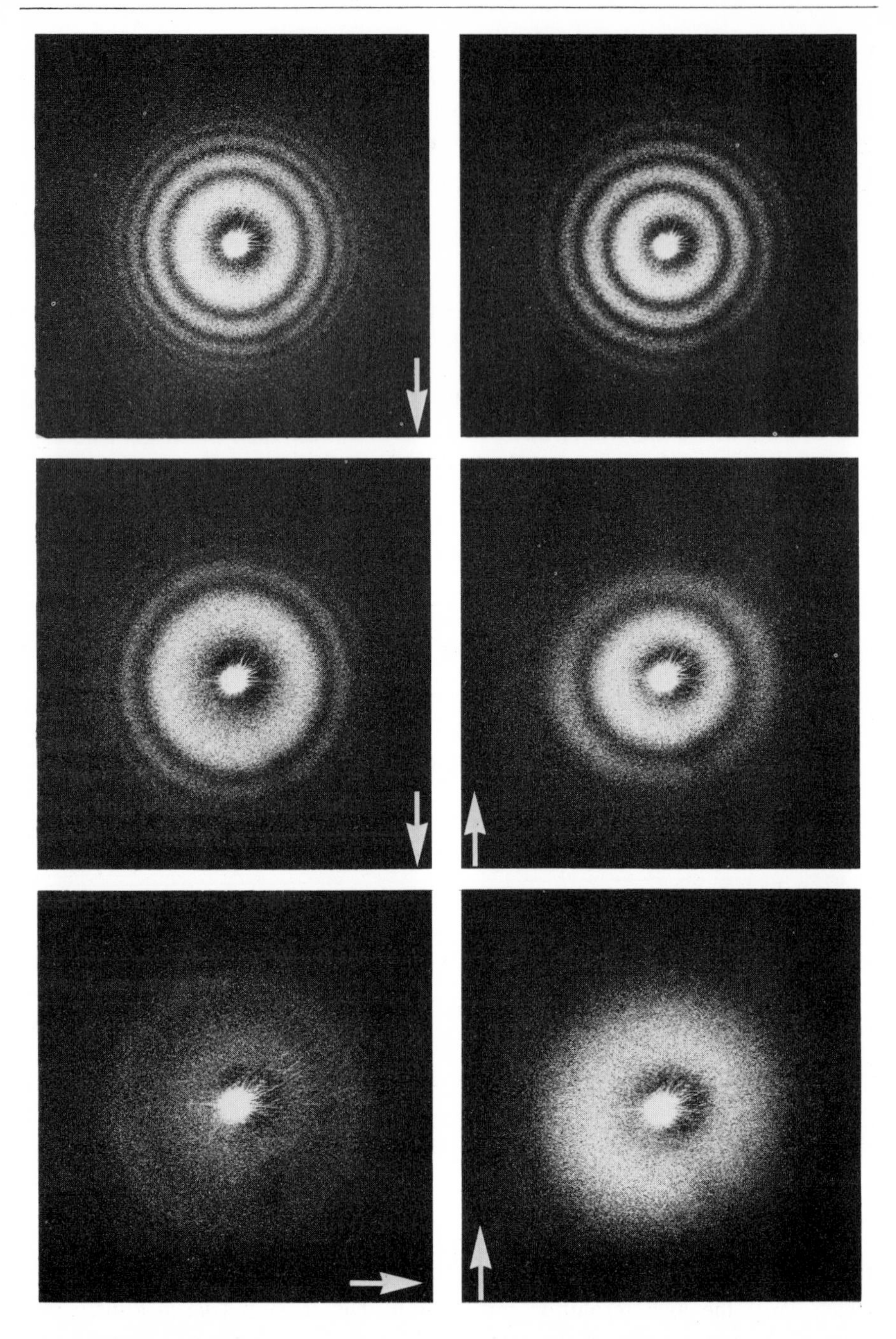

Fig. 26. Diffraction patterns of a typical focal series of a carbon film. 'Over-focus' at top left. Arrows indicate direction of decreasing lens current. Gaussian focus – bottom left. Scherzer focus – bottom right.

to reproduce fine detail in the specimen. Fig. 26 shows a typical focal series of six successive micrographs of a carbon support film starting at 'overfocus' at the top left. Following the arrows which indicate decreasing objective lens current, the third pattern is the diffraction pattern at approximately 'in' focus or Gaussian focus (low contrast image, not recommended for picture taking). The fourth pattern is the optimum (or Scherzer) underfocus and the diffraction pattern shows a continuous bright disc surrounding the bright central spot. This indicates a broad pass band of spatial frequencies, all of the same sign, i.e. with no contrast reversal. On these diffraction patterns, of course, the finest detail occurs at the largest visible radius. The axial symmetry of this set of diffraction patterns indicates that the objective lens is transmitting fine detail equally, regardless of its orientation. This means that the lens astigmatism has been adequately corrected. Since the eye is very sensitive to axial symmetry, much smaller amounts of astigmatism can be detected by this method than is possible using Fresnel fringes.

It should also be mentioned that the Scherzer focus position cannot be recognized by eye on the microscope screen. A useful technique is to find the Gaussian focus first by noting the point of minimum image contrast and then counting how many fine-focus clicks below this focus gives the corresponding Scherzer diffraction pattern in a through-focal series. If the objective lens is focused too far under focus, a set of bright and dark concentric circles appears in the diffraction pattern indicating that the image contrast reverses from positive to negative for different spatial frequencies, since adjacent bright circles in the pattern are in fact reversed in contrast. The dark circles correspond to spatial frequencies which are not being transmitted or imaged at all. Since most TEM samples are mounted on carbon support films, it is usually possible, with the aid of the diffractometer, to deduce much interesting information about the state of the instrument from routine micrographs. It is an interesting fact that in laboratories equipped with optical diffractometers, micrographs are often rejected purely on the basis of a screening by the optical diffractometer even when they appear satisfactory under a 10 × magnifier. For example, slight astigmatism can often give rise to false structures in the micrograph, which are extremely difficult to distinguish by eye from the true specimen detail, since the eye has no means of detecting which details have suffered contrast reversal.

The effect of objective lens astigmatism is shown in Fig. 27. The perfect circles of Fig. 26 have become ellipses whose long axes, for example, turn through 90° (Maltese Cross pattern) on passing through focus. This set of patterns represents fairly severe astigmatism, typical of fault conditions in the specimen stage or objective region of the microscope.

Figure 28 shows the effect of reducing the size of the objective aperture. It should be remembered that it is the outer zones of the objective aperture which carry the high resolution information. The same is true in a diffraction pattern, the outer zones of the pattern indicating the presence of high resolution information. Thus, Fig. 28(a) shows a diffraction pattern taken with a large objective aperture not filled by the imaging beam of the microscope; when the

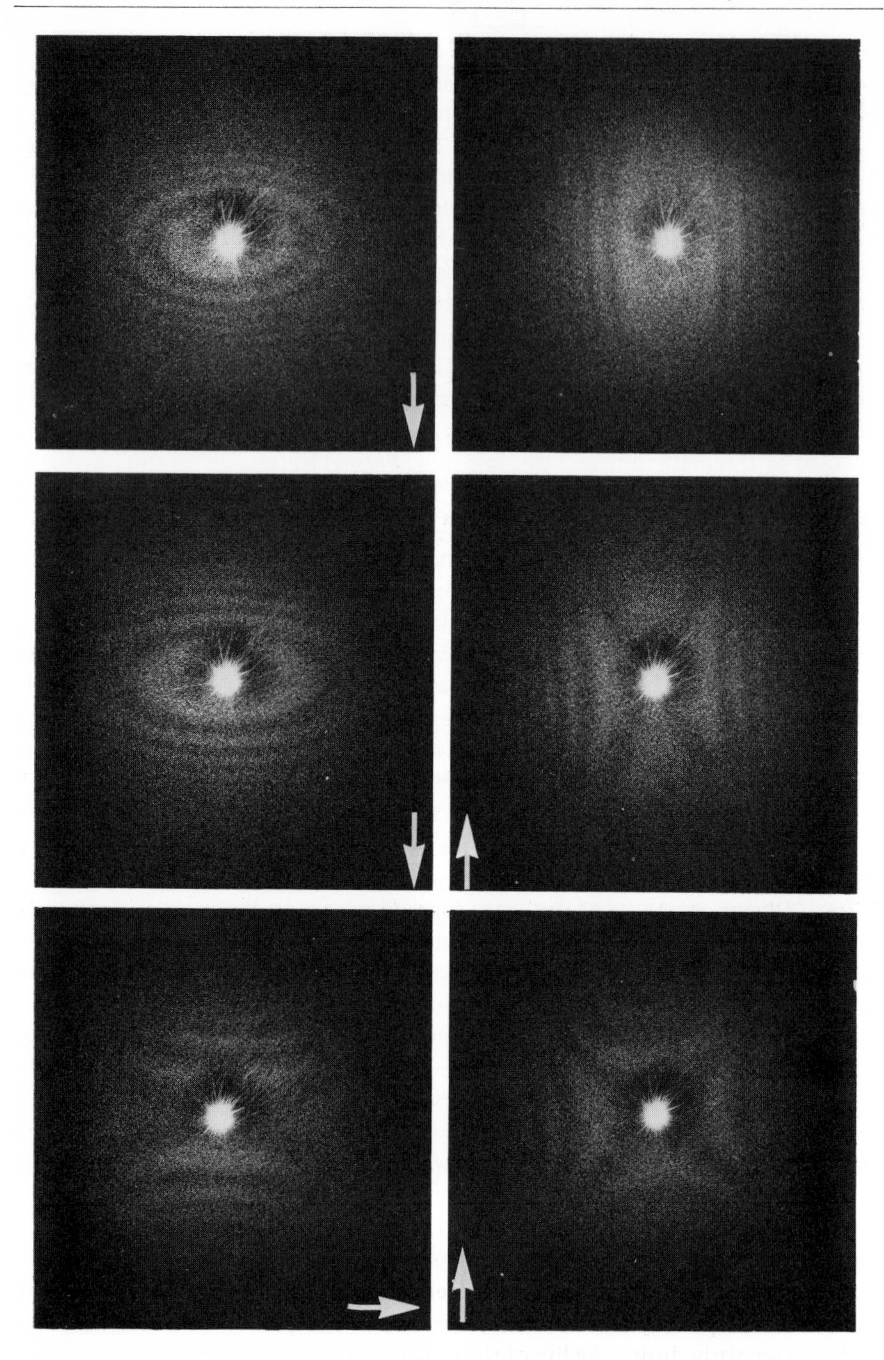

Fig. 27. Focal series corresponding to that of Fig. 26 but with severe astigmatism.

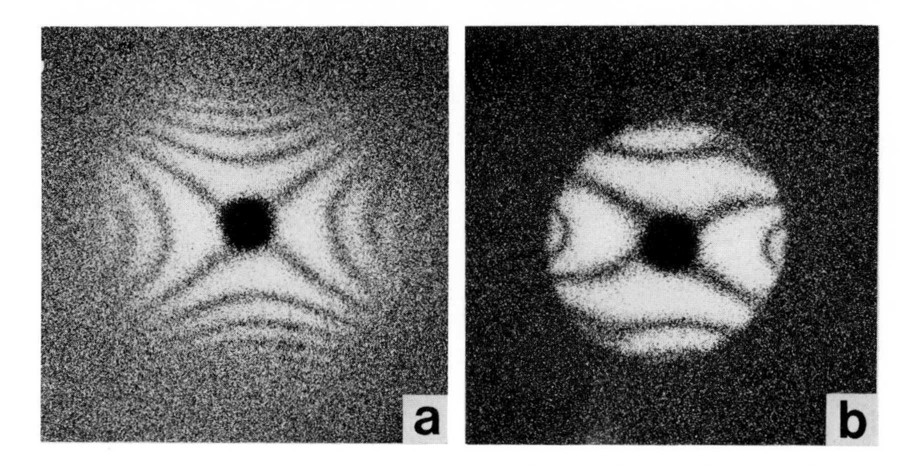

Fig. 28. Effect on diffraction pattern of reducing the objective aperture. (a) Large aperture (b) small aperture, cutting off high spatial frequencies.

objective aperture is reduced in size, Fig. 28(b), the outer edges of the pattern are stopped out and only the central part of the pattern is preserved. When the objective aperture is made even smaller only the lower resolution information is transmitted, though the picture often looks 'better' because of improved contrast. This is a very useful way of checking the actual numerical aperture of the TEM and of diagnosing malfunctioning of the objective lens.

Specimen stage drift is particularly difficult to diagnose by eye especially in the presence of astigmatism. Here the optical diffractometer can often help. Figure 29(a) shows the effect of stage drift, or vibration, on an otherwise circular diffraction pattern. The pattern is 'wiped out' in the direction at right angles to the line of drift. Figures 29(b) and (c) show the same specimen drift but now in the presence of varying degrees of astigmatism in the objective lens.

Finally, Fig. 30 shows circular diffraction patterns from a carbon film supporting a crystalline specimen. From the uniformly intense pattern from the carbon film we may deduce that Fig. 30(a) corresponds to the ideal Scherzer focus, whereas Fig. 30(b) corresponds to a focal setting below Scherzer focus. In this micrograph the same diffraction spots are weakened or lost in the zeroes of the image transfer function (dark zones of the diffraction pattern). In addition, specimen information corresonding to the outer bright zone will be reversed in contrast and will therefore give rise to interpretation problems in the micrograph.

A simplified optical diffractometer permanently available will soon justify its modest cost, especially if it is installed before the acceptance tests on a new TEM and used regularly during the life of the instrument.

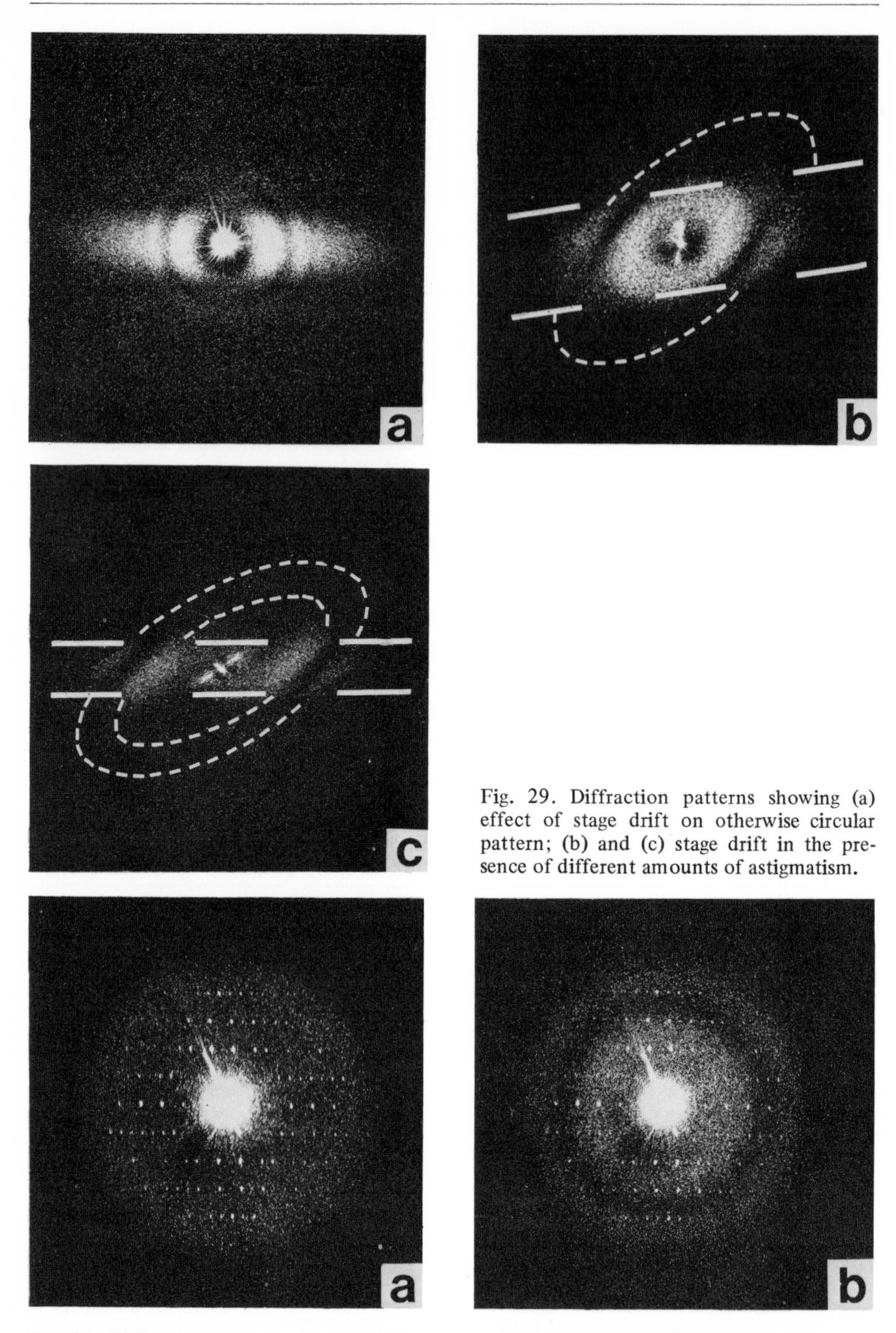

Fig. 29. Diffraction patterns showing (a) effect of stage drift on otherwise circular pattern; (b) and (c) stage drift in the presence of different amounts of astigmatism.

Fig. 30. Diffraction patterns showing diffraction spots from a crystalline specimen on a carbon film.
(a) At Scherzer focus.
(b) Below Scherzer focus (see text).

Appendices

1. Examples of standard values (= knob position)

Chapter reference *Example*

1.1 Emission current meter – dark current at each kV
20–*0*, 40–*1*, 60–*2*, 80–*3*, 100–*5*
Standing current
20–*8*, 40–*36*, 60–*55*, 80–*74*, 100–*95*

1.2 Bias and filament controls with a new filament

1.3 Lens current values – minimum and maximum (amps) 100 kV at
C1 *1–4*, C2 *0.5–3*, O *2–5*, D *0.1–2*, P1 *0.1–2*, P2 *0.5–3*

1.4 Condenser lens/high voltage performance at 100 kV
C1 *2 amps* C2 *1.2 amps*
or *10 steps* or
or *5 μm spot size*

1.7 Objective lens/high voltage performance focus at 100 kV
Coarse Medium Fine Ultra fine
or *3.1 amps*

1.8 Objective lens stigmator at 100 kV
X: *3.1 turns* or , Y: *5.7 turns* or

1.9 Diffraction–intermediate lens/high voltage performance at 100 kV
Lens: *1.4 amps* or

1.10 Imaging lens combinations in amps at 20 kV

Mag.	O	D	P1	P2
1 K	*1.68*	*0.25*	*off*	*0.5*
3 K	*1.68*	*0.25*	*off*	*0.85*
5 K	*1.7*	*0.48*	*0.28*	*1.02*
10 K	*1.7*	*0.52*	*0.22*	*1.34*
20 K	*1.7*	*0.68*	*0.19*	*1.34*
50 K	*1.7*	*0.58*	*0.62*	*1.34*
100 K	*1.7*	*0.58*	*0.85*	*1.34*

(Voltage readings could be used in place of the above)

1.10 Magnification calibration check at critical points on Fig. 3 at 100 kV.
2500 × *low 2%*, 25 000 × *low 3%*, 250 000 × *low 5%*

1.13 Reference points at 100 kV.

Voltages	5	7.5	12	24	48
	4.9	*7.45*	*11.98*	*23.8*	*47.94*

Stabilities	HV	C1	C2	O	D	P1	P2
+ or –	*2*	*1*	*1.5*	*0.5*	*1*	*0.5*	*1*

1.14 Vacuum system

Test 1 to 4 at 5×10^{-2} mbar: 5 at 10^{-4} mbar: 6 at 10^{-3} mbar

1	2	3	4	5	6
44 s	*15 s*	*95 s*	*80 s*	*21 min*	*19 min*

2.3 Resolution at 100 kV

2.4 After 1 hour *0.6 nm* After 3 hours *0.45 nm*

2.5 Drift rate at 100 kV *0.025 nm* s^{-1}

2.6 Contamination rate

Without LN *0.09 nm* s^{-1}, With LN *0.00083 nm* s^{-1}

at 100 kV Emission 20 Spot 3 C ap 150 0 ap 30 films 4

2.8 Camera length

kV	P1	P2	L
100	*0.6 amp*	*1.8 amp*	*12.8 cm*

2. Cleaning record example

Filament number	53			54		
Fitted	*4:1:85*			*18:2:85*		
Failed	*18:2*			*12:4*		
Life	*85 h*			*94 h*		
Type of break	*Normal*			*Normal*		
Cathode cleaned	*4:1*			*16:2*		
Anode cleaned	–			*19:3*		
Chamber and insulator	–			*19:3*		
Specimen holders	*4:1*	*25:1*	*15:2*	*8:3*	*29:3*	*12:4*
Cold finger	*4:1*	*1:2*		*1:3*	*4:4*	
Apertures changed	*4:1*			*1:3*		
Holders cleaned	–			*1:3*		
Special notes	*4 users*			*2 users*		

3. Trouble shooting flow charts

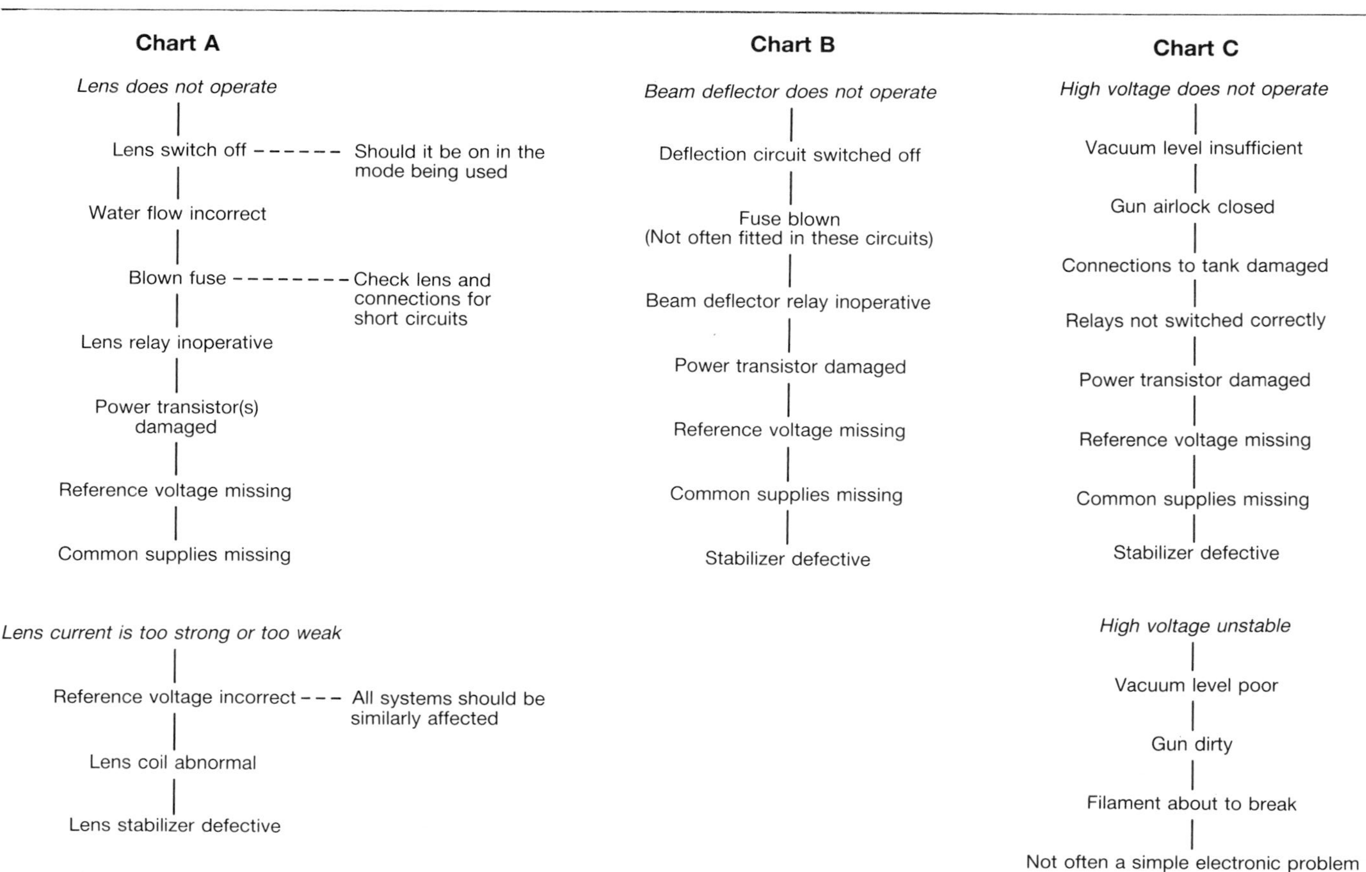

Index